RHEOLOGY OF WHEAT PRODUCTS

Edited by Hamed Faridi

Published by

**The American Association of Cereal Chemists, Inc.
St. Paul, Minnesota**

COVER: Illustration adapted from a construction of a structural relaxation curve charting dough resistance. The original graph appears on page 152.

This publication is based on presentations from a symposium entitled "Rheology of Wheat Products" held on September 24, 1985, in conjunction with the annual meeting of the American Association of Cereal Chemists in Orlando, FL. To make the information available in a timely and economical fashion, this book has been reproduced directly from typewritten copy submitted in final form to the American Association of Cereal Chemists by the editor of the volume. No editing or proofing has been done by the Association.

Reference in this publication to a trademark, proprietary product, or company name by personnel of the U.S. Department of Agriculture is intended for explicit description only and does not imply approval or recommendation to the exclusion of others that may be suitable.

Library of Congress Card Number: 85-073386
International Standard Book Number: 0-913250-42-2

©1985 by the American Association of Cereal Chemists

All rights reserved. No part of this book may be reproduced in any form by photocopy, microfilm, retrieval system, or any other means, without written permission from the publisher.

Printed in the United States of America

The American Association of Cereal Chemists
3340 Pilot Knob Road
St. Paul, Minnesota 55121, USA

CONTRIBUTORS

W. BUSHUK, Department of Plant Science, University of Manitoba, Winnipeg, Manitoba, Canada

B.L. D'APPOLONIA, Department of Cereal Science and Food Technology, North Dakota State University, Fargo, North Dakota

J.W. DICK, Department of Cereal Science and Food Technology, North Dakota State University, Fargo, North Dakota

K.C. DIEHL, Department of Agricultural Engineering, Virginia Polytechnic Institute and State University, Blacksburg, Virginia

P.C. DREESE, Department of Grain Science and Industry, Kansas State University, Blacksburg, Virginia

P.W. RUSSELL EGGITT, Dalgety U.K. Ltd., Group Research Laboratory, Cambridge, United Kingdom

J.M. FAUBION, Department of Grain Science and Industry, Kansas State University, Manhattan, Kansas

C.S. FITCHETT, Dalgety U.K. Ltd., Group Research Laboratory, Cambridge, United Kingdom

P.J. FRAZIER, Dalgety U.K. Ltd., Group Research Laboratory, Cambridge, United Kingdom

G.R. FULCHER, Agriculture Canada, Ottawa Research Station, Ontario, Canada

K.M. HARDY, Department of Food Science, University of Guelph, Guelph, Ontario, Canada

R.C. HOSENEY, Department of Grain Science and Industry, Kansas State University, Manhattan, Kansas

R.H. KILBORN, Canadian Grain Commission, Grain Research Laboratory, Winnipeg, Manitoba, Canada

W.H. KUNERTH, USDA/ARS Wheat Quality Laboratory, Fargo, North Dakota

J. LOH, Central Research Division, General Foods Corporation, Tarrytown, New York

A. MILLER, Flour Milling and Baking Association, Chorleywood, Herts, United Kingdom

S. NAGAO, Research Center, Nisshin Flour Milling Co., Ltd., Ohi-machi, Saitama, Japan

J.G. PONTE, JR., Department of Grain Science and Industry, Kansas State University, Manhattan Kansas

K.R. PRESTON, Canadian Grain Commission, Grain Research Laboratory, Winnipeg, Manitoba, Canada

V.F. RASPER, Department of Food Science, University of Guelph, Guelph, Ontario, Canada

PREFACE

Rheology can be defined as the study of the deformation and flow of matter. The production of wheat-based products from flour, water, and other ingredients is a process in which the rheological properties of the material change considerably at various processing stages. Measurement of the rheological properties give valuable information concerning the quality of the raw materials, the machining properties of the dough and the textural characteristics of finished products.

The complexity of the both components in wheat (proteins, starch, lipids, pentosans, etc.), and the macromolecular structure of dough, complicate physical dough testing methods. There are two basic types of rheological measurements; the fundamental and the empirical. In the fundamental, test results are expressed in terms of basic physical quantities such as stress, strain, or rate of strain. The results therefore are independent of instrument or operator. Empirical tests are usually more easily performed and the results are correlated with flour performance in the bakery.

Because bakeries have become large and automated, flour uniformity is desirable and accurate prediction of variations in handling characteristics help avoid production-down-time and finished product rejects. Better understanding of mixing, fermentation, and machining processes and means to measure and control these processes can enormously affect finished product quality. Loss of textural properties of baked goods due to aging needs to be determined objectively. Therefore, it is no surprise that currently there is a growing interest in various aspects of rheology of wheat-based products.

This volume is a collection of 12 papers presented at the Rheology of Wheat Products Symposium at the 70th Annual Convention of American Association of Cereal Chemists in Orlando, Florida, September 22-26, 1985. The volume presents an objective, state-of-the-art discussion of various aspects of wheat products rheology.

My sincere thanks to the speakers at the symposium and to the authors who contributed manuscripts for this proceeding and to the staff of the AACC National Office for their editorial assistance.

I dedicate this volume to all cereal rheologists, worldwide, whose endeavors may make this important science more and more applied to every aspect of baking technology.

>   Hamed Faridi
>   Nabisco Brands, Inc.
>   Technology Center
>   East Hanover, NJ

CONTENTS

	Page
Rheology: Theory and application to wheat flour doughs. W. BUSHUK	1
Use of the Mixograph and Farinograph in wheat quality evaluation. W.H. KUNERTH and B.L. D'APPOLONIA	27
Constant water content vs. constant consistency techniques in alveography of soft wheat flours. V.F. RASPER, K.M. HARDY and G.R. FULCHER	51
Do-corder and its application in dough rheology. SEIICHI NAGAO	75
Dynamic rheological testing of wheat flour doughs. J.M. FAUBION, P.C. DREESE and K.C. DIEHL	91
The use of a penetrometer to measure the consistency of short doughs. A.R. MILLER	117
Grain Research Laboratory instrumentation for studying the breadmaking process. R.H. KILBORN and K.R. PRESTON	133
Laboratory measurement of dough development. P.J. FRAZIER, C.S. FITCHETT and P.W. RUSSELL EGGITT	151
Rheology of fermenting dough. R.C. HOSENEY	177
Rheology of soft wheat products. J. LOH	193
Rheology of durum. J.W. DICK	219
Rheology of bread crumb. J.G. PONTE, JR. and J.M. FAUBION	241

RHEOLOGY OF WHEAT PRODUCTS

RHEOLOGY: THEORY AND APPLICATION TO WHEAT FLOUR DOUGHS[1]

W. Bushuk
Department of Plant Science
University of Manitoba
Winnipeg, MB   R3T 2N2

INTRODUCTION

Rheology is the science that deals with deformation of matter. Deformation can be one of flow and hence involve the properties of viscosity or consistency, or it can be an elastic or plastic deformation. In rheology, we are not concerned with movements of complete bodies but in the movements of parts of bodies relative to each other. Movements of bodies as a whole is the subject of mechanics, a branch of physics. The difference between rheology and mechanics can be illustrated by considering the properties of three common substances - water, rubber, and jam. Water flows and hence has the property of viscosity; rubber is elastic and can be characterized by a modulus of elasticity; and jam is plastic and can be characterized by viscosity. Under the action of external forces, these three substances will behave differently and it is this behavior that is of concern in rheology. The three substances would behave similarly, if as in mechanics, their

---

[1]Contribution No. 724

movements as a whole were considered. The pertinent property in this case is the mass of the whole body.

Rheological changes in substances result from the action of various forces. If the deformation reaches a certain level due to an applied force but returns to its initial state when the force is removed, then it is an elastic deformation. If the deformation remains permanently when the forces are removed, then it is a plastic deformation. In liquids, the deformation increases continually under the action of a finite force and the material flows.

The main interest in fundamental rheology derives from the interrelationships between molecular structure of materials and their rheological properties in solid or solution state. In applied science or technology, rheology additionally provides information that is useful in control of raw materials and process vis a vis the quality of the final product. In this regard rheological behavior can be used as a guide in designing machines for industrial processes.

In breadmaking, the dough undergoes some type of deformation in every phase of the process. During mixing, dough undergoes extreme deformations beyond the rupture limits; during fermentation the deformations are much smaller; during sheeting and shaping, deformations are of an intermediate level; and finally during proofing and baking, dough is subjected to more deformations. Accordingly, the application of rheological concepts to the behavior of dough seems a natural requirement of research on the interrelationship among flour composition, added ingredients, process parameters and the characteristics of the loaf of bread.

THEORY

Before we can consider dough rheology *per se*, it is necessary to introduce a number of concepts and certain terminology that are fundamental to the science of rheology.

Deformation of materials results from the application of a force or a load. The S.I. unit of force is the newton (N) which is defined as the force which will give an acceleration of 1 metre per second to a mass of 1 kg. One newton equals 10^5 dynes. Stress (traction in some texts), the state of a body under the action of a force, is best defined as the force per unit area.

If a force acts on a body or it is in a state of stress, it undergoes deformation such as extension and compression. A more useful term is relative deformation – the change in a dimension relative to the original dimension – or strain.

Deformation in liquids is called flow and since flow is a rate process it is best expressed as a rate of strain. The science of rheology attempts to provide equations of state giving mathematical relationships between stress and strain or rate of strain.

First, let us consider the property of viscosity. Viscosity of a fluid is that property which determines the resistance to motion when a shearing force is exerted on a fluid under laminar flow. Turbulent flow which results when the forces are extremely large, is excluded from this definition. The coefficient of dynamic viscosity of a fluid, η, is defined as the tangential force on a unit area of either of two parallel planes at a unit distance apart when the space between the planes is filled with the fluid and one of the planes moves relative to the other with unit velocity in its own plane. Mathematically this can be stated as,

$$\eta = \frac{S/d\gamma}{dt} \qquad (1)$$

where S = stress or applied force per unit area, and

$\frac{d\gamma}{dt}$ = rate of strain or deformation.

The dimensions of the viscosity coefficient are $ML^{-1}T^{-1}$ which in the S.I. system is Nsm^{-2}. If a shear force of 1 newton produces a flow of 1 metre per second then the coefficient of viscosity, η, equal one poiseuille (10 poise) or $1Nsm^{-2}$.

Equation (1) shows that for an ideally viscous system, the stress is directly proportional to the rate of deformation. This is Newton's Law and systems that obey this law are called Newtonian. For non-Newtonian systems, the coefficient of viscosity depends on the rate of deformation. The true coefficient for such systems must be obtained by extrapolation to zero rate of deformation.

Ordinary (non-ideal) materials exhibit four general types of non-Newtonian behavior. These are classified into time-independent and time-dependent behavior. Time-independent non-Newtonian behavior includes fluids that undergo thinning (decrease of coefficient of viscosity) with increasing rates of shear (pseudoplastic) and those that become thicker with increasing rates of shear (dilatant). Time-dependent behavior includes those substances that show a decrease in coefficient of viscosity with time (at constant rate of shear), e.g. thixotropic and those that show an increase in coefficient of viscosity with time e.g. rheopectic. For non-Newtonian substances, the coefficient of viscosity at any point in the flow curve is an "apparent" coefficient. The "true" coefficient is obtained by extrapolating the flow curve to zero rate of shear or zero time. Furthermore, because of the time-dependency, flow curves for thixotropic and rheopectic substances show a "hysteresis" effect.

Figure 1 illustrates the viscous behavior described above in terms of standard flow curves.

Another basic property of rheological systems that is important in dough rheology is elasticity. Since an elastic deformation is one which disappears entirely when the force which caused it is released, it is therefore a fundamental property of solids.

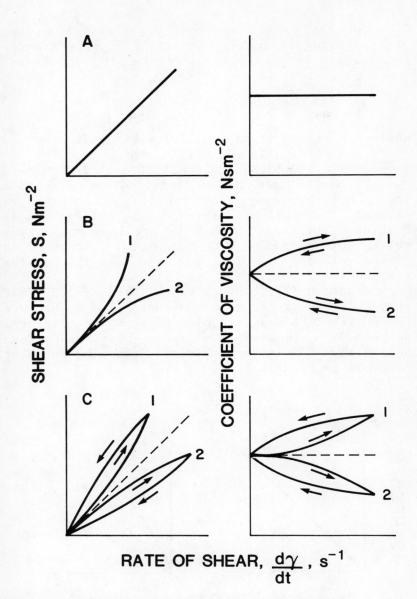

Figure 1. Flow curves: A, Newtonian liquids; B-1, non-Newtonian, dilatant; B-2, non-Newtonian, pseudoplastic; C-1, non-Newtonian, rheopectic; and C-2, non-Newtonian, thixotropic.

For many solid systems, especially when the applied force is small, the stress (applied force) is directly proportional to deformation or strain. For such systems we can write,

$$S = G\gamma \quad (2)$$

where S = stress or applied force per unit area
γ = deformation or strain and
G = proportionality constant called modulus.

If the strain is in longitudinal direction (one dimension caused by tensile or compressive stress), the modulus in this case is called Young's Modulus. The units of modulus are the same as the units of stress (ML^{-1}) since strain is dimensionless. The S.I. units of Young's Modulus are Nm^{-2}.

Equation (2) is Hooke's Law. The systems that obey this law are Hookean and those that do not are non-Hookean. A standard plot of equation (2) for a Hookean body is shown in Figure 2A. Non-Hookean bodies show non-linear behavior, depending on shear rate, as in Figure 2B.

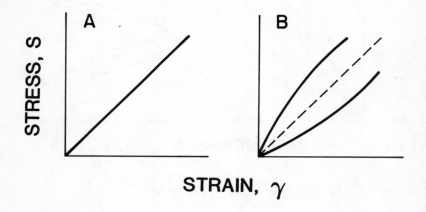

Figure 2. Stress-strain curves for Hookean (A) and non-Hookean (B) solids.

Experimentally, elastic modulus is usually obtained by measuring the strain resulting from a known applied stress. The experiment must be performed in such a way that Hooke's Law is obeyed. This means that the deformation must be completely reversible. If the recovery is instantaneous after the applied force is removed, the elasticity is considered ideal, and if recovery is complete but requires a finite time, then we have retarded elasticity. If, as in Figure 2A, the stress is plotted against strain, the slope of the curve is the elastic modulus.

The rheological approach that has been applied to doughs with some success is based on a combination of the viscosity and elasticity theories. Accordingly, to appreciate dough rheology it is necessary to consider a viscoelastic system. Such systems are referred to as Maxwell, Kelvin-Voigt, or Burger body, depending on how the viscous and elastic elements are combined, and are intermediate in rheological properties between viscous liquids and elastic solids.

For a viscoelastic system, the viscous and elastic contributions can be regarded as additive. From equation (1) we have the rate of viscous strain given by

$$\left(\frac{d\gamma}{dt}\right)_\eta = \frac{S}{\eta} \qquad (3)$$

while by differentiating equation (2) the rate of elastic strain given by

$$\left(\frac{d\gamma}{dt}\right)_G = \frac{1}{G}\frac{dS}{dt} \qquad (4)$$

The total or the viscoelastic strain can be obtained by adding equations (3) and (4).

$$\frac{d\gamma}{dt} = \frac{1}{G}\frac{dS}{dt} + \frac{S}{\eta} \qquad (5)$$

In rheology, the ratio of the viscosity coefficient and the elastic modulus (η/G) is defined as the relaxation time and is represented by the symbol τ. Introducing the relaxation time into equation (5), we obtain the fundamental differential equation that describes a viscoelastic system undergoing a deformation due to an applied stress.

$$G \frac{d\gamma}{dt} = \frac{dS}{dt} + \frac{S}{\tau} \qquad (6)$$

Solutions to equation (6) can be obtained by imposing boundary conditions applicable to specific instruments or testing procedures. Some of the more successful attempts to determine the fundamental parameters viscosity coefficient, elastic modulus and relaxation time for doughs will now be discussed.

RHEOLOGICAL PROPERTIES OF DOUGH

In attempting to apply the fundamental concepts of rheology as outlined above to bread dough, it is necessary first of all to ask the question whether this will serve any real useful purpose. The answer of course is an unqualified yes; however this is not generally appreciated by the practical cereal chemist concerned with problems in the mill or in the bakery (see Hibberd and Parker, 1975 for review).

First of all, in applying fundamental rheological considerations to dough, we are attempting to describe a highly complex system in terms of meaningful parameters such as viscosity, elastic modulus, and relaxation time. When these parameters are determined for a particular dough, it should then be theoretically possible to describe the behavior of this dough for a different set of experimental conditions. Another important value of these parameters is that they provide the researcher with information on the structure of dough, and to the practical mill or bakery chemist a means for setting up flour and ingredient specifications. The section that follows reviews the attemps made to examine dough on the basis of theoretical concepts discussed.

The first systematic attempt to determine fundamental rheological parameters for dough were made by Schofield and Scott Blair in the 1930's (1932, 1933a, b, 1937). In one of their studies (1937), they placed samples of dough on a mercury bath and stretched them by means of a winch. The stress was then released and the dough allowed to undergo elastic recovery. The extension against time curve that they obtained is shown in Figure 3. They divided the strain (calculated from the extension) into two parts. The irrecoverable part of the strain was considered the viscous (plastic) flow, and the recoverable part was the elastic deformation. Knowing the rate of deformation, $d\gamma/dt$, and the applied force or stress, they were able to calculate the coefficient of viscosity. The values thus obtained for various doughs ranged from 0.55×10^6 to 15×10^6 poise. From the applied force and the resulting elastic deformation, they obtained values for elastic modulus ranging from $(1.6 \text{ to } 5.5) \times 10^4$ c.g.s. units.

The work of Schofield and Scott Blair was subsequently extended to a variety of different doughs and experimental conditions by Halton and Scott Blair (1937). Their values for η and G agreed with those of Schofield and Scott Blair (1937).

Later Muller <u>et al</u> (1961,1962) succeeded in obtaining the coefficient of viscosity and elastic modulus for various doughs from measurements with the Brabender Extensigraph. These workers first of all calibrated the extensigraph in c.g.s. units. The rate of extension was expressed in cm. s^{-1}, and the applied force or stress in dyne cm^{-2}.

To separate the viscous and elastic components, Muller and co-workers followed the method of Schofield and Scott Blair (1933). The

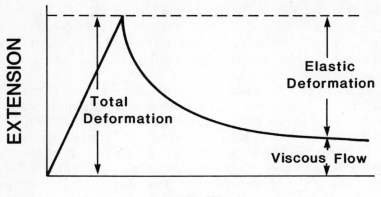

Figure 3. Strain versus time curve for dough on application and release of stress.

dough piece was stretched to a given extension and the applied force then removed to permit elastic recovery. Again, viscous deformation was the irreversible part and the elastic deformation was the reversible part. From these experiments, Muller et al (1962) obtained values for η ranging from (0.3 to 3.6) x 10^4 poise and for G ranging from (0.5 to 6.7) x 10^4 c.g.s. units. Their values for G agreed well with those of Schofield and Scott Blair (1933), however the values for η were two orders of magnitude lower.

The calculations discussed above assumed that dough is both a Newtonian liquid and a Hookean solid. However this is a gross oversimplification. The non-linear rheological behavior of dough was demonstrated by Scott Blair and co-workers (1932,1933a,b,1937), accordingly the values of the coefficient of viscosity and elastic modulus are valid only for the particular conditions of the experiment and hence have little value as fundamental parameters. Furthermore their practical value is limited also since they apply to a narrow range of experimental conditions and hence only to a very small part of the breadmaking process. In modern processes, dough is subjected to a wide range of stresses; from extremely high in the high speed developers of the continuous processes to extremely low during fermentation and proofing. A more fruitful rheological approach might be to focus attention on a single phase of the breadmaking process where the deformations to which the dough is subjected are relatively constant and can be duplicated experimentally.

As already indicated, a dough can be considered as a composite body having both viscous and elastic properties. To such a system we can apply equation (6). This equation can be tested experimentally by solving the equation for boundary conditions that are applicable to particular instruments. A number of the more successful of these approaches will now be discussed.

Let us first consider the case where the dough is deformed at constant rate of strain as in the Brabender Extensigraph. Since $d\gamma/dt$ is a constant, equation (6) simplifies to

$$\frac{S}{\eta} = R(1 - e^{-t/\tau}) \qquad (7)$$

Where S = applied stress
R = a constant (rate of strain)
η = coefficient of viscosity
t = time and
τ = relaxation time.

The graph of equation (7) is shown in Figure 4A. Figure 4B shows the more familiar extensigram. Although the boundary condition of equation (7) prevails in the extensigraph, this instrument cannot be used directly to test this equation. The main difficulty is that it is not possible to determine the dimensions of the test piece throughout the extension process. Approximations of cross section can be made from the length of the dough piece, but these are highly inaccurate. If we examine the extensigram in Figure 4B, relative to the plot of equation (7) (Figure 4A), it is apparent that the experimental curve is a complex function of a number of variables. Qualitatively we can say that the shape of the extensigram depends on three main factors: (1) the coefficient of viscosity, (2) the relaxation time, and (3) the continuous decrease in cross-sectional area of the test piece throughout the extension.

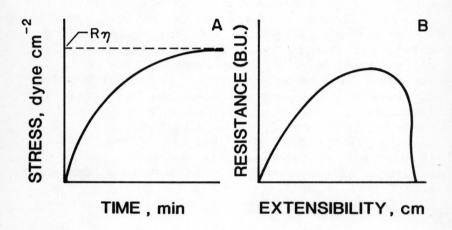

Figure 4. Plots of fundamental equation (7) and a typical extensigraph for a wheat four dough.

To obtain η and τ from the extensigram, it is necessary to account for the third factor. Although a number of attempts have been made, it has not been possible to derive the viscous and elastic components from the normal extensigram. Nevertheless the Brabender Extensigraph has provided much useful information on dough properties.
The difficulties of the Extensigraph can be obviated by studying the viscoelastic behavior of dough by measuring the relaxation of stress by a specially designed relaxometer. This approach was applied to dough with considerable success by Hlynka and co-workers (Hlynka and Anderson, 1952;

Cunningham et al, 1953; Cunningham and Hlynka, 1954).

In stress relaxation, the dough sample is first stretched to a predetermined extension and the tension is allowed to decrease at constant strain. For this experimental arrangement, equation (6) can be solved to give

$$S = S_o e^{-t/\tau} \qquad (8)$$

where S = stress at time t
S_o = stress at $t = 0$, and
τ = relaxation time.

Stress relaxation measurements lead to families of stress relaxation curves for related experimental conditions. Examples of such curves are shown in Figure 5. From these curves, it is possible to obtain stress relaxation time which can then be used to evaluate the coefficient of viscosity and the modulus of elasticity. Following this procedure Cunningham and Hlynka (1954) obtained 2.5×10^6 poise for the viscosity coefficient and $(0.6 - 0.7) \times 10^4$ c.g.s. units for the elastic modulus. These values agreed in general with those of Schofield and Scott Blair (1933).

Another approach to the consideration of dough as a viscoelastic substance is to apply the condition of constant stress. For such conditions, equation (6) gives the following function for the deformation:

$$\gamma = \frac{S}{G} (1 + e^{-t/\tau}) \qquad (9)$$

Equation (9) can be handled experimentally by a creep experiment. This type of experiment was applied to dough by Glucklich and Shelef (1962) and Bloksma (1962). The latter objected to the use of existing instruments, namely the Chopin Alveograph and the Brabender Extensigraph for three reasons:

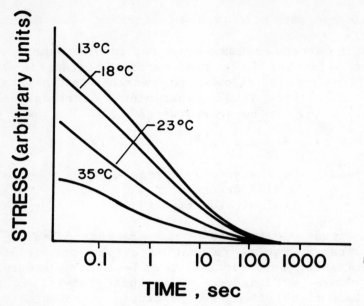

Figure 5. Stress relaxation curves for wheat flour doughs at different temperatures (reproduced from Cunningham and Hlynka, 1954)

1. The deformation of the test piece in the two instruments is not homogeneous; some parts of the dough deform more and faster than others.
2. When the deformation is not homogeneous, both strain and stress are complex functions of time.
3. The experimental rates of deformation in the two instruments are much higher than the rates in a fermenting dough.

Bloksma (1962) therefore adapted the truncated cone and plate viscometer in which the applied stress was held constant and the deformation was similar to that occurring in dough during fermentation. With this instrument he obtained values for modulus of elasticity of about 10^4 dynes cm^{-2}, which agree quite well with previously published values. Bloksma also (1962) introduced to dough rheology a new parameter, shear compliance, which he defined as the ratio of the shear strain and the shear stress. For a particular shear stress and time, shear compliance is the reciprocal of the elastic modulus. The compliance values varied from 10^{-5} to 10^{-3} g.$^{-1}$ cm.s^2 and depended on shear stress, water absorption, and the grade of the flour. Bloksma's study confirmed earlier observations of Scott Blair and co-workers that the rheological behavior of dough is non-linear. He found that the compliance increased by a factor of over 10 when the shear stress increased from 200 to 5000 dyne cm^{-2}. Doughs exhibited some linearity at very low and high shear stresses where the behavior was either predominantly elastic or viscous. Since Bloksma's study was designed to study the rheological behavior during fermentation, it is not directly applicable to the mixing phase.

Shimizu and Ichiba (1958) used a dynamic method to study fundamental rheological properties of dough. These workers developed an apparatus in which the dough sample was held in a fixed outer cup and was in contact with a suspended oscillating bob. From the angle between the torque on the bob and its motion, these workers obtained values of $(0.2 - 2) \times 10^4$ for the coefficient of viscosity and $(0.5 - 2) \times 10^4$ for the modulus of elasticity. The values for the viscosity coefficient are two orders of magnitude lower than previously published values. Bloksma and Hlynka (1960) suggested that the agreement might have been better if the frequency range was extended to lower values.

Two new approaches to fundamental dough rheology were reported in the 1960's. Hibberd (Hibberd and Wallace, 1966; Hibberd, 1970a,b) designed an instrument in which the dough was sinusoidally sheared in the annulus between a fixed outer cylinder and an oscillating inner cylinder. This instrument was applied to a relatively wide range of shear strains but still well below the rupture limit of dough. It was shown that the linear viscoelastic theory applies only to strains smaller than 4×10^{-3}. This finding confirmed the view that most of previously published data derived from viscoelastic measurements were based on an incorrect assumption that linearity persists to much higher strains. The study of Hibberd is also interesting from the practical point of view in that an attempt was made to analyze the varying stresses imposed on the dough during various phases of the breadmaking process.

Another interesting study is that of Rinde et al (1967) who developed a procedure for measuring the stress-strain and rupture behavior of doughs under extremely large deformations. Dough test pieces were in the form of a ring which was looped and stretched over Teflon^R hooks at constant strain rates while immersed in a liquid of the same density. For each extension rate, the effect of continuous stress relaxation could be separated from the overall non-linear stress-strain relation. These workers succeeded in expressing the stress-strain behavior of a dough over a fairly wide range of time, temperature and water absorption by a single equation based on a characteristic modulus, and time, temperature, and water absorption factors. In relation to baking quality, it was found that the strain at rupture was less for a poor quality flour than for a good quality flour at higher temperature, but the order reversed for lower temperatures. The stress at rupture was four times greater for the higher quality flour for all conditions investigated. This study is a significant step forward in the application of fundamental rheological measurements obtained with a number of different empirical instruments.

DOUGH RHEOLOGY, CHEMISTRY AND PROCESSING

This section will review briefly those aspects of dough chemistry and processing that contribute significantly to those rheological properties of doughs that are important to commercial breadmaking. For a more comprehensive treatment of this subject, the reader is referred to the excellent review by Bloksma (1971).

There is good evidence that the rheological properties of dough derive directly from analogous properties of gluten. In this context, gluten is considered to comprise mainly of protein. Recently, increasing emphasis has be placed on the contribution of non-protein flour constituents, especially lipids an starch. Reports from several laboratories (see Bushuk et al, 1984 for review) present convincing evidence that specific non-protein constituents of flour interact strongly with specific proteins of gluten. Accordingly one would expect that such interactions would contribute significantly to the rheological properties of the resulting gluten. So far, this contribution has been described in a qualitative way only; quantitation of the interactions in terms of fundamental rheological parameters such as viscosity and modulus of elasticity remains for future research. Detailed discussion of the contribution of interactions to dough rheology is beyond the scope of this article.

Rheological properties of gluten depend on its main proteins, gliadin and glutenin. Gliadin, which comprises single chain molecules ranging in molecular weight from about 30,000 to 80,000, forms a highly viscous mass when mixed with about the same proportion of water as that used to form a dough. Accordingly, it is assumed that this constituent contributes the property of viscosity to gluten. On the other hand, glutenin comprises a highly heterogeneous mixture of proteins in terms of molecular weight and structure. In addition to single polypeptide-chain molecules, glutenin contains many different molecules composed of several polypeptide chains (subunits) crosslinked with disulfide (S-S) bonds. Such molecules can differ in several ways; molecular weight, molecular structure, and subunit composition. All of these aspects would contribute to the rheological properties of the substance. Isolated glutenin, on hydration, forms a highly elastic mass and for this reason it is presumed to contribute the elastic component to the viscoelastic properties of gluten and dough. For a comprehensive treatment of the interrelationship between molecular properties of glutenin and its contribution to the rheological properties of doughs, the reader is referred to the series of publications by Ewart (1985 and references therein).

It was mentioned earlier that intermolecular interactions, between gluten proteins and between protein and nonprotein constituents contributre significantly to the rheological properties of dough. Presumably these interactions lead to the formation of various aggregates (films, fibrils and micelles) whose contributions to rheological properties of dough would depend on their structure and tendency to interact with other aggregates.

The most important interactions appear to result from hydrogen bonds. Evidence for their contribution to dough properties has been obtained from experiments on the effects of heavy water (D_2O) and urea, a well known hydrogen bond breaker. The fact that almost 90% of glutamic acid, which forms more than 30% of the gluten protein, is in the form of the amide, is generally cited as evidence of the presence extensive hydrogen bonding in gluten. Similarly, hydrophobic interactions can be stronly implicated in the physical properties of doughs. Electrostatic interactions (salt linkages) also make a positive contribution but magnitude would be small because of the relatively low content of amino acids in gluten proteins with side chains which would ionize at normal pH of doughs.

All of the secondary cross linkages in dough are weak individually but collectively they form strong interactions. However, these interactions would be extremely "mobile" when the stability provided by the covalent S-S cross linkages is removed.

The chemical groups in gluten proteins that

have received much attention are the sulfhydryl (-SH) and disulfide (S-S) groups (for review see Bloksma, 1975). The former because they react with oxidizing dough improvers and the latter because they provide stability to dough by immobilizing parts of protein chains and thereby facilitating formation of hydrogen bonds and hydrophobic interactions. Furthermore S-S groups through their ability to interchange under stress are probably involved in mechanical dough development and in the relaxation of strains induced into doughs during mixing, molding, and other processing steps where the dough is subjected to physical stress. The importance of these two chemical groups to physical properties of dough can be readily demonstrated by instruments such as the farinograph and extensigraph. Various aspects of the role of -SH and S-S groups in doughs have been extensively investigated but so far there is no single theory that explains all of the experimental facts (Frater et al, 1960 and references therein).

The effects of increasing or decreasing the number of -SH and/or S-S groups in dough are strongly inter-related with the physical manipulations that are applied to doughs during various stages of the breadmaking process. For quick and efficient dough development during the mixing stage, the dough must have some "mobile or rheologically active" -SH groups (Jones et al, 1974). If all of the -SH groups are removed by reaction with fast acting oxidizing agents such as potassium iodate or blocking agents such as N-ethylmaleimide the dough cannot be "structurally activated" during rounding and shaping for the extensigraph test. It is presumed that during structural activation (or mechanical development), the mobile -SH groups reduce "rheologically active" S-S groups and thereby initiate S-S interchanges which facilitate re-orientation of protein chains required for the formation of a "developed" dough structure. Reduction of stabilizing S-S groups also facilitates exchange of hydrogen bonds and other physical interactions. By a combination of S-S interchange and the physical action of mixing, sheeting and moulding, the structure of gluten is developed into a structure that provides optimum gas retention and crumb structure in the loaf of bread. The -SH groups are only needed during the mixing stage. For optimum baking quality they must subsequently be eliminated otherwise they will continue to promote S-S interchanges; the dough will relax and lose its developed structure. This effect is particularly important in the early oven phase of the breadmaking process where the dough is subjected to extremely high stresses. So, for optimum involvement of the -SH and S-S system in dough, it is necessary to have a high concentration of -SH during the mixing stage but the concentration must be drastically decreased if the dough is to maintain its stability just prior to the setting of the loaf structure in the oven, when the dough is subjected to extremely high stresses. This is why ascorbic acid or a combination of cysteine and potassium bromate are such excellent dough improvers. Both provide fast reducing actions and slow oxidizing actions. As is well known, the balance of reducing and oxidizing actions of dough improvers is extremely critical in "no-time" doughs where much of the development is achieved by the mixing actions.

Direct involvement of -SH groups in "structural activation" (as for example, increase in resistance to extension in the extensograph test produced by rounding and shaping) and subsequent relaxation of the induced strains was first reported by Frater et al (1961). Results demonstrating these effects are shown in Figure 6. The effect of the elimination of -SH groups can be demonstrated by adding a fast acting oxidizing agent such as potassium iodate. N-ethylmaleimide, a fast-acting -SH blocking agent produces the same effect as iodate in this type of experiment. On the basis of results such as shown in Figure 6, it has been concluded that the key effect of oxidizing improvers is the removal of -SH groups and not the formation of S-S bonds as had been thought earlier.

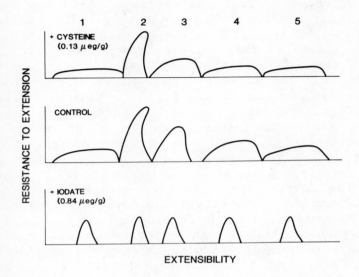

Figure 6. Extensigrams of dough containing varying amounts of -SH stretched at different time intervals after rounding and shaping. All doughs were rounded and shaped immediately after mixing. Doughs No. 1 were stretched after 45 min rest and then re-rounded and stretched after resting for 20 s (No. 2), 5 min (No. 3), and 15 min (No. 4).

In this final section attention will be focussed on effects of mechanical work on rheological properties of dough in so far as it is possible to isolate these effects from those of chemical groups of flour constituents and added ingredients. Relative to the discussion that follows, it should be kept in mind that the observed changes generally result from a strong interaction between chemical and processing factors.

During mixing of flour and water to form a dough, the dough undergoes a physical transformation generally referred to as "development". Dough development is used to describe desirable changes that eventually result in optimal baking performance of a specific flour. In the context of this article, development usually means the optimization of the viscoelastic properties. For any one flour, mixer and type of bread, optimum development is achieved by a specific mixing time. As indicated earlier, many of the flour constituents and added ingredients interact strongly with the mixing action. In regard to added ingredients, the effects of atmospheric oxygen, which is continuously mixed into the dough must not be overlooked. An experienced baker can detect the point at which the dough has reached optimum development by feeling it or stretching between the fingers. Instruments that measure work input during mixing and provide a mixing curve have replaced the baker's "finger" in modern automated bakeries. These instruments,along with the traditional farinograph and extensograph are used to optimize formulations in terms of water absorption and addition of reducing and oxidizing dough improvers.

For a more detailed discussion of the changes in doughs that result from mechanical work and the recovery of original properties with resting (relaxation), the reader is referred to articles by Bloksma and Hlynka (1960) and Dempster et al (1952).

The effect of mechanical work is generally explained on the basis of re-orientation of interacting constituents, facilitation of S-S interchanges and, under some conditions (when all -SH groups are oxidized or blocked) by actual breakage of S-S bonds (Hoseney 1985). The observed effects can be explained in a qualitative way but more research is needed to develop an all-inclusive mathematical treatment of the measured rheological changes.

In relation to mechanical development, there is substantial evidence that indicates that developed doughs are extremely "fragile" and can be easily "undeveloped (unmixed)" by subsequent mechanical work (Paredes-Lopez and Bushuk 1982, Tipples and Kilborn 1975). During undevelopment by slow mixing, a developed dough changes in character and takes on the appearance of an undermixed dough. If such doughs are baked, the resulting bread is lower in loaf volume and poorer in grain structure. The characteristics of a developed dough can be recovered by remixing the undeveloped dough at the appropriate speed. An analogous observation, which probably involves the same rheological changes, is the redevelopment of overmixed doughs by remixing after a short period of relaxation (Bloksma and Hlynka 1960).

There are numerous opportunities for the modification of the rheological properties of doughs by appropriate combination of chemical and physical variables. Further application of rheological concepts to the improvement of baking technology remains an interesting challenge for future dough rheologists.

ACKNOWLEDGEMENT

I am indebted to the late Dr. I. Hlynka of the Grain Research Laboratory whose publications and lectures aroused my interest in dough rheology and in no small way form the basis of this review.

LITERATURE CITED

BLOKSMA, A.H. 1962. Slow creep of wheat flour doughs. Rheol. Acta 2:217.
BLOKSMA, A. H. 1971. Rheology and chemistry of dough. Page 523 in: Wheat Chemistry and Technology. Y. Pomeranz, ed. American Association of Cereal Chemists, Inc., St. Paul.

BLOKSMA, A.H. 1975. Thiol and disulfide groups in dough rheology. Cereal Chem. 52:170r.

BLOKSMA, A.H., and HLYNKA, I. 1960. The effect of remixing on the structural relaxation of unleavened dough. Cereal Chem. 37:352

BUSHUK, W., BEKES, F., McMASTER, G.J., and ZAWISTOWSKA, U. 1984. Carbohydrate and lipid complexes with gliadin and glutenin. Pages 101-109 in: Gluten Proteins: Proceedings of the 2nd International Workshop on Gluten Proteins, A. Graveland and J.H.E. Moonen, eds., TNO Wageningen, The Netherlands, May 1-3

CUNNINGHAM, J. R. and HLYNKA, I. 1954. Relaxation time spectrum of dough and the influence of temperature, rest and water content. J. Appl. Phys. 25:1075.

CUNNINGHAM, J. R., HLYNKA, I., and ANDERSON, J. A. 1953. An improved relaxometer for viscoelastic substances applied to the study of wheat dough. Can. J. Technol. 31:98.

DEMPSTER, C.J., HLYNKA, I., and WINKLER, C.A. 1952. Quantitative extensograph studies of relaxation of internal stresses in non-fermenting bromated and unbromated doughs. Cereal Chem 29:39.

EWART, J.A.D. 1985. Blocked thiols in glutenin and protein quality. J. Sci. Food Agric. 36:101.

FRATER, R., HIRD, F.J.R., and MOSS, H.J. 1961. Role of disulphide exchange reactions in the relaxation of strains introduced in dough. J. Sci. Food Agric. 4:269.

FRATER, R., HIRD, F.J.R., MOSS, H. J., and YATES, J.R. 1960. A role for thiol and disulphide groups in determining the rheological properties of dough made from wheaten flour. Nature, Lond., 186:451.

GLUCKLICH, J., and SHELEF, L. 1962. An investigation into the rheological properties of flour dough. Studies in shear and compression. Cereal Chem 39:242.

HALTON, P., and SCOTT BLAIR, G.W. 1936. A study of some physical properties of flour doughs in relation to their bread-making qualities. J. Phys. Chem. 40:561.

HIBBERD, G.E. 1970. Dynamic viscoelastic behaviour of wheat flour doughts. II. Effects of water absorption in the linear region. Rheol. Acta 9:497.

HIBBERD, G.E. 1970. Dynamic viscoelastic behaviour of wheat flour doughs. III. The influence of the starch granules. Rheol. Acta 9:501.

HIBBERD, G.E. and PARKER, N.S. 1975. Measurement of the fundamental rheological properties of wheat-flour doughs. Cereal Chem. 52: 1r.

HIBBERD, G. E. and WALLACE, W. J. 1966. Dynamic viscoelastic behaviour of wheat flour doughs. I. Linear aspects. Rheol. Acta 5:193.

HLYNKA, I., and ANDERSON, J.A. 1952. Relaxation of tension in stretched dough. Can. J. Technol. 30:198.

HOSENEY, R.C. 1985. The mixing phenomenon. Cereal Foods World 30:453.

JONES, I.K., PHILLIPS, J.W., and HIRD, F.J.R. 1974. The estimation of rheologically important thiol and disulphide groups in dough. J. Sci. Food Agric. 25:1.

MULLER, H.G., WILLIMAS, M.V., RUSSELL EGGITT, P.W., and COPPOCK, J.B.M. 1961. Fundamental studies on dough with the Brabender extensograph. I. Determination of stress-strain curves. J. Sci. Food Agr. 12:513.

MULLER, H.G., WILLIAMS, M.V., RUSSELL EGGITT, P.W., and COPPOCK, J.B.M. 1962. Fundamental studies on dough with the Brabender extensograph. II. Determination of the apparent elastic modulus and coefficient of viscosity of wheat flour dough. J. Sci. Food Agr. 13:572.

PARADES-LOPEZ, O., and BUSHUK, W. 1983. Development and "undevelopment" of wheat dough by mixing: Physicochemical studies. Cereal Chem. 60:19.

RINDE, J.A., SMITH, T.L., and TSCHOEGL, N.W. 1967. Stress-strain and rupture behavior of wheat flour doughs under large tensile deformations.

Program 52nd Annual Meeting, A.A.C.C., p. 50.
SCHOFIELD, R.K., and SCOTT BLAIR, G.W. 1932. The relationship between viscosity, elasticity and plastic strength of soft materials as illustrated by some mechanical properties of flour doughs. I. Proc. Roy. Soc. (London) A138:707.
SCHOFIELD, R.K., and SCOTT BLAIR, G.W. 1933a. The relationship between viscosity, elasticity and plastic strength of soft materials as illustrated by some mechanical properties of flour doughs. II. Proc. Roy. Soc. (London) A139:557.
SCHOFIELD, R.K., and SCOTT BLAIR, G.W. 1933b. The relationship between viscosity, elasticity and plastic strength of soft materials as illustrated by some mechanical properties of flour doughs. III. Proc. Roy. Soc. (London). A141:72.
SCHOFIELD, R.K., and SCOTT BLAIR, G.W. 1937. The relationship between viscosity, elasticity and plastic strength of soft materials as illustrated by some mechanical properties of flour doughs. IV. The separate contributions of gluten and starch. Proc. Roy. Soc. (London) A160:87.
SHIMIZU, T., and ICHIBA, A. 1958. Rheological studies of wheat flour dough. I. Measurement of dynamic viscoelasticity. Bull. Agr. Chem. Soc. Japan 22:294.
TIPPLES, K.H., and KILBORN, R.H. 1975. "Unmixing" - The disorientation of developed bread doughs by slow speed mixing. Cereal Chem 52:248.

USE OF THE MIXOGRAPH AND FARINOGRAPH IN WHEAT QUALITY EVALUATION[1]

W.H. Kunerth
USDA/ARS Wheat Quality Laboratory
Fargo, ND 58105

B.L. D'Appolonia
Dept. of Cereal Science and Food Technology
North Dakota State University
Fargo, ND 58105

INTRODUCTION

Two of the most widely used physical dough testing instruments for wheat quality evaluation studies are the mixograph and farinograph. Both of these instruments are of the dynamic type since they perform measurements on a dough as the dough is being mixed. However, the design of each instrument, including the type of mixing action, is quite different and although both instruments produce a tracing or curve as the dough is being mixed the derived information may be somewhat different.

The purpose of this paper is to present some of the basic principles involved with both instruments, their advantages and disadvantages and their utilization for wheat quality studies.

Farinograph

The most complete and comprehensive source of information dealing with the farinograph is the 3rd edition of The Farinograph Handbook (D'Appolonia and

---

[1] This article is in the public domain and not copyrightable. It may be freely reprinted with customary crediting of the source: The American Association of Cereal Chemists, Inc. Mention of firm names or products does not constitute endorsement by the USDA over others of similar nature.

Kunerth 1984). The nine chapters in this handbook include, among others, discussions on the operation and theoretical aspects of the farinograph, types and interpretation of farinograms, uses and new developments of the instrument, as well as precautionary measures to be taken when using the instrument.

Figure 1 shows the basic parts of a farinograph. Of major importance is the mixing bowl which comes in the 50 or 300 g size. The mixing action is brought about by two sigma-type blades which rotate at a differential speed of 3:2. The type of mixing created by this type of blade is different than that exerted by the pin-type mixer. Temperature during mixing is controlled by the use of temperature-controlled water circulating in a jacket surrounding the bowl. The driving mechanism and force for the mixing blades are furnished by a dynamometer illustrated in Figure 1.

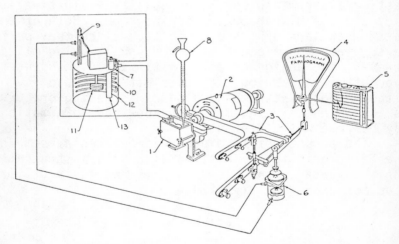

Figure 1. Basic parts of a farinograph: mixing bowl (1), dynamometer (2), lever system (3), scale system (4), recording mechanism (5), dashpot (6), thermostat (7), buret (8), thermoregulator (9), cooling coils (10), circulating pump (11), reservoir tank (12), heating element (13). (Diagram adapted from a figure supplied by C. W. Brabender Instruments, Inc. Figure taken from The Farinograph Handbook).

Although farinographs are equipped to operate at two speeds, the test is normally performed at the fast speed, 62 rpm on the slow paddle and 93 rpm on the fast paddle. For very weak flours the slow speed, 31 rpm on the slow paddle and 46.5 rpm on the fast paddle is used. Occasionally in flour quality evaluation two peaks are obtained in the farinogram. The two-peak type farinogram is often produced by flours that contain a very strong or "bucky"-type of gluten. The first peak may not reflect gluten development but rather is a function of hydration. The two-peak phenomenon may be altered by increasing the speed of the two paddles. The time to reach the second peak correlates much better with the mixing time in a bakery than does the initial peak.

In a breeding program, sample size may be limited, hence a 50 g bowl is often more practical than a 300 g bowl. However, it has been demonstrated that the farinograms obtained using the 300 g bowl are somewhat stronger than those obtained from corresponding flours using the 50 g bowl.

Information derived from a typical farinogram is shown in Figure 2. This includes peak time, stability, mechanical or mixing tolerance index (MTI), valorimeter value, arrival time, departure time and time to breakdown. The measurements most used are probably peak time and either stability or MTI values. These measurements indicate the amount of work input required to develop a dough to optimum development and how much additional mixing can be imparted to the dough before it begins to break down.

Besides the farinogram measurements one additional important piece of information obtained from this instrument is farinograph absorption. This value gives the amount of water that can be added to a flour so that the dough has a fixed consistency. Although the baking absorption may not necessarily be the same as the farinograph absorption, in many cases there is a good correlation between the two. An example in which farinograph absorption does not relate to baking absorption is in flour samples containing relatively high levels of starch damage.

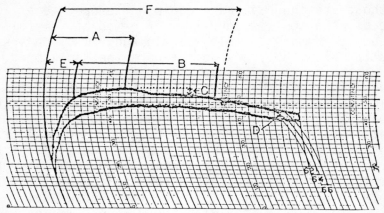

Figure 2. Readings commonly used in interpreting a farinogram: peak time (A), stability (B), mechanical tolerance index (C), valorimeter value (D), arrival time (E), departure time (E + B), time to breakdown (F). (Figure taken from The Farinograph Handbook).

With such samples the farinograph absorption will be higher than the corresponding baking absorption. With high starch damage samples the water held by the damaged starch may be released during fermentation as the starch is degraded by the amylase enzymes. Such doughs after fermentation will be wet and sticky indicating the necessity for a reduction in absorption.

The type or shape of the farinogram curve obtained varies according to the variety, class of wheat and environmental conditions as well as the grade or type of flour produced during the milling operation. Hard wheat or bread wheat type flours will usually produce farinograms that have much longer stability values than flours derived from soft wheats.

Besides the effect of flour protein content on the farinogram curve, protein quality is equally important. Table I illustrates the effect of variety on farinogram properties. One notes from this table that the hard red spring (HRS) wheat varieties Olaf and James, have similar protein content yet have extremely different farinograph properties. The curves for these two varieties are shown in Figure 3.

TABLE I
Difference in Farinograms for Different
Spring Wheat Varieties Grown
at the Same Location

Variety	Flour Protein[a] (%)	Flour Absorption[a] (%)	Peak Time (min)	Mixing Stability (min)
Olaf	13.8	65.2	24.5	29.0
James	13.9	61.9	8.0	11.0

[a]On 14.0% moisture basis

The arrival time for different flours can vary considerably. An extremely short arrival time indicates that hydration takes place rapidly whereas a long arrival time indicates that the water is being taken up by the various components present in the flour at a much slower rate. The relative water uptake of the major dough components was estimated by Bushuk (1966) to be 0.44 g/g of granular starch, 2.0 g/g of damaged starch, 2.2 g/g of protein and 15 g/g of pentosan. Differences in the degree of damaged starch and in protein quantity and quality are the most likely factors to affect the rate of hydration.

Figure 4 shows the farinograms of four grades of flour from the same wheat. The differences shown reflect the importance of the type of milling process used to obtain flour for wheat quality evaluation studies.

High protein bread wheat flours are very often used to blend with lower protein, lower quality wheat flours indigenous to many countries that import high quality wheat. Figure 5 shows the effect on farinograms of incorporating a HRS wheat flour in 10% increments with a soft red winter (SRW) wheat flour. The effects noted will, of course, depend upon the quality of the HRS and SRW wheat flours.

The farinograph is a fairly universal physical

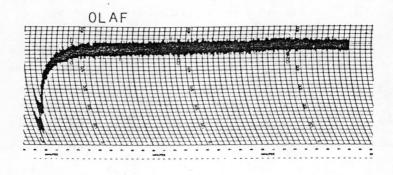

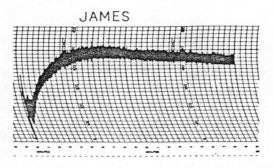

Figure 3. Farinograms of two spring wheats of similar protein content grown at one location, showing effect of variety.

dough testing tool used throughout the world in cereal testing laboratories. The instrument itself has been used for many different purposes, the most important of which probably has been as a wheat quality control tool. In the HRS wheat quality testing program of the Department of Cereal Science and Food Technology at North Dakota State University, the farinograph is used in conjunction with several other dough testing instruments. The primary pieces of information derived from the farinograph in our program are absorption and overall mixing properties of the samples under evaluation. The data reported include absorption, peak time, stability and mixing tolerance index. In addition, we have devised a

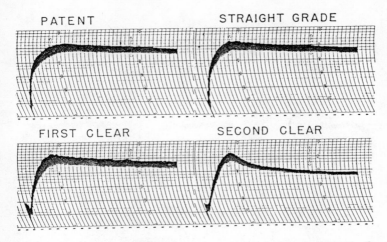

Figure 4. Farinograms of four grades of flour from the same wheat. (Figure taken from The Farinograph Handbook).

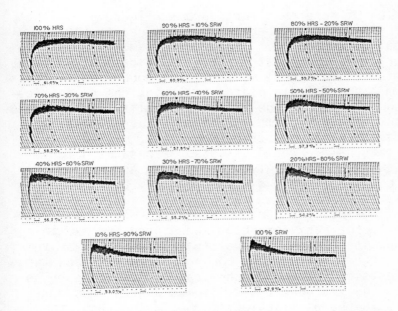

Figure 5. Farinograms for blends of HRS and SRW wheat flours. (Figure taken from The Farinograph Handbook).

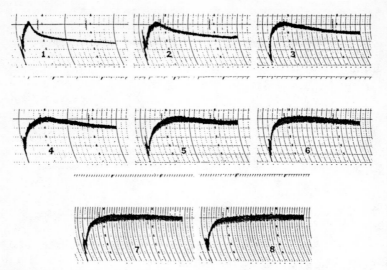

Figure 6. Reference farinograms for HRS wheat.

series of reference farinograms (Figure 6) so that a classification number can be assigned to any particular farinogram. If a farinogram has curve characteristics that differ dramatically from those presented in Figure 6, such as a double peak, the term "abnormal" is placed following the classification number. In the HRS wheat quality evaluation of plant breeders' samples, if a farinogram has a classification number of 4 or less, mixing properties of that sample would be rated as unsatisfactory. Likewise, abnormal type farinogram curves or excessively strong gluten types may be rated as unsatisfactory.

Besides its use as a quality control tool for wheat, the farinograph has been used for examining composite flours, reconstituted type flour systems, the effects of various baking ingredients and the effects of various biochemical components such as proteins, carbohydrates, lipids and enzymes.

In the utilization of the farinograph, the technique of the operator is of extreme importance in obtaining meaningful and reproducible results. Any factor that affects absorption has a corresponding effect on other farinogram values. Consequently incorrect measurement of flour weight or water added will have an effect on absorption and the resultant shape of the curve. Needless to say, proper operation of the instrument is essential. The mixing bowl and blades, balance and lever systems, and chart recorder must all be such that the resultant farinogram is representative of the particular sample under evaluation.

Various studies have been conducted in which farinograph data have been compared with other instruments and procedures used to evaluate flour. Early studies by Landis and Freilich (1935) showed that the farinograph can be used for standardizing the dough development rate of various mixers.

In our laboratories the farinograph has been used to evaluate different series of samples. Among these different series are HRS wheat experimentals and named cultivars in our wheat breeding programs, and commercial samples of HRS wheat collected during the new crop harvest.

Table II shows the correlation coefficients obtained between the three farinograph properties, peak time, tolerance and MTI, and wheat and flour protein content for two separate studies. Study 1 is based on the evaluation of 107 wheat samples collected at crop harvest over a five-year period. Each of the 107 samples was a composite of a number of samples collected from different districts within each of the four states of North and South Dakota, Minnesota and Montana. Highly significant correlations were obtained between either wheat or flour protein content and each of the three farinogram parameters. For study 2, 240 samples were utilized. These samples consisted of ten cultivars grown at six locations for four crop years. Once again the correlations were highly significant at the 1.0% level. Highly significant correlations were also obtained in both studies between the same farinogram measurements and wet gluten content.

TABLE II
Correlation Coefficients of Pertinent Farinogram
Measurements With Protein Content

	Farinogram Parameter		
	Peak Time	Tolerance	MTI
Study 1[a]			
Wheat Protein	0.515**	0.535**	-0.478**
Flour Protein	0.517**	0.487**	-0.443**
Study 2[b]			
Wheat Protein	0.401**	0.430**	-0.500**
Flour Protein	0.416**	0.448**	-0.513**

[a] $N = 107$ samples. All values highly significant at the 1.0% level.
[b] $N = 240$ samples. All values highly significant at the 1.0% level.

Table III shows the correlation coefficients obtained for the data collected in studies 1 and 2 (Table II) between the farinogram measurements and loaf volume. In study 1 no significant correlations were obtained between loaf volume and farinogram measurements while in study 2 highly significant correlations were found. In study 2 the samples evaluated were pure varieties while in study 1 the samples were composites of commercially grown wheat samples within the different districts of the state. There were probably greater differences in quality in the samples used for study 2, because of the use of different cultivars rather than mixtures of cultivars. For study 2 correlation coefficients were also obtained between the farinograph measurements and baking mixing time. The results indicated that the farinograph does provide useful information con-

TABLE III
Correlation Coefficients of Pertinent Farinogram
Measurements With Loaf Volume

	Farinogram Parameter		
	Peak Time	Tolerance	MTI
Study 1[a]			
Loaf Volume	0.124	0.093	-0.098
Study 2[b]			
Loaf Volume	0.321**	0.252**	-0.340**

[a] N = 107 samples.
[b] N = 240 samples. All values highly significant at the 1.0% level.

cerning mixing properties but by itself cannot be used to establish overall quality. In a quality evaluation study of 240 samples which involved ten cultivars grown at six locations for four crop years, the correlations obtained between baking mixing time and farinogram peak time, tolerance and MTI were 0.24258, 0.26668 and -0.25007, respectively. Although the correlations were low, they were all highly significant.

MIXOGRAPH

The mixograph is a small, high speed recording dough mixer formerly manufactured by National Manufacturing Company of Lincoln, Nebraska and presently available from TMCO, 501 J Street, Lincoln, Nebraska. It was originally designed by Swanson and Working (1933) to provide a method of measuring quality as far as quality is related to gluten structure. The device measures the rate of dough development, the maximum resistance of the dough to mixing, and the duration of resistance to mechanical overmixing.

The five basic parts of the mixograph are shown in Figure 7. The mixer is available in two sizes requiring either 10 g or 25-35 g flour.

The mixing action of the mixograph is provided by four vertical planetary pins revolving about three stationary pins in the bottom of the bowl. The mixing action can be described as a pull, fold and repull action which is much more severe than is produced by the farinograph. As a result, the main advantage of the mixograph is the speed with which a test can be conducted. As the dough is developed, an increasing force is required to force the revolving pins through the dough. This increasing force is measured as a tendency to rotate the bowl which is mounted on the center of a lever system. The resistance to rotation is provided by a tension spring having twelve possible tension settings which allow adjustment of the tension to a level suitable for the type of flour being tested. Positions eight to eleven are most commonly used. Higher settings are used for comparisons between doughs of strong flours and lower settings for weak flours.

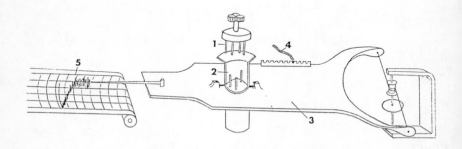

Figure 7. The basic parts of the mixograph: mixing pins (1), mixing bowl (2), swivel base or lever system (3), tension spring (4) and kymograph (5).

The characteristics of the mixogram curve are the result of the changing plastic, elastic, and viscous properties of the dough during mixing. During the initial ascent of the curve, water is being brought into contact and absorbed by the protein and starch, and because of the folding and stretching action of the mixing pins, the dough begins to be developed. As the dough develops, the force required to move the pins through the dough increases to a maximum plasticity or minimum mobility. This point corresponds to the top of the curve. Beyond this point mechanical degradation of the dough causes an increase in mobility resulting in the curve sloping downward and tailing off.

Many attempts have been made to assign numerical values to various parameters of mixograph curves and to relate these measurements to quality factors (Johnson et al 1943, Swanson and Johnson 1943, Morris et al 1944, Sibbitt et al 1953, Shuey 1974). Some of these measurements are shown in Figure 8 and are identified as follows:

Peak time. The time required for the peak to reach maximum height and is similar to the dough-development time of the farinograph (line DC). An increase in peak time is associated with an increase in mixing time required for development of the dough.

Area under the curve. This area is enclosed by the base line and a line drawn through the center of the curve until either a specified time has elapsed since the mixograph was started or a specified time has elapsed since the curve peak was attained. The larger the area the stronger the flour and the greater its tolerance to overmixing.

Peak height. The height from the base line to the center of the curve at maximum plasticity is the peak height (line CH). The measurement provides information about flour strength and absorption.

Height of a curve at a specified time after peak or start of mixing. The height of the curve at a specified time is similar to the farinograph "tolerance index" or "drop off" values. Higher values indicate a flour which is more tolerant to mixing.

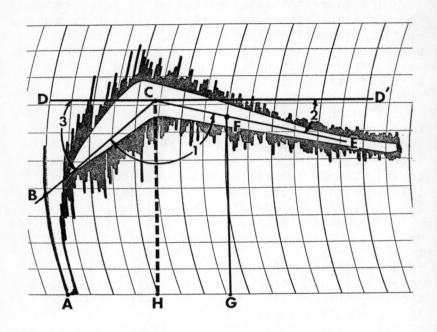

Figure 8. Mixogram measurements.

Angle between ascending and descending portions of the curve at peak. This angle is obtained by drawing a line from the center of the curve at its peak down the center of the curve in both directions (angle 1). A large angle is associated with a more tolerant flour.

Weakening angle. This angle is formed by drawing a line from the center of the curve at its peak down the descending portion of the curve and a line horizontal to the base line through the center of the curve at its maximum height (angle 2). The size of the angle is inversely related to mixing tolerance.

Development angle. This angle is formed by a line drawn horizontal to the base line through the center of the curve at its maximum height and a line drawn through the center of ascending portion of the curve (angle 3).

The shape of mixograph curves vary according to wheat class and variety as well as the environmental conditions under which the wheat was grown. Figure 9 shows the variability in mixogram characteristics between five different classes of wheat, and Figure 10 shows the variability between four varieties of hard red spring wheat.

Larmour et al (1939) used the mixograph in comparing protein content and baking quality of hard red winter wheats. The curves obtained revealed certain distinctions between varieties as well as within varieties. Certain varieties exhibited such distinctive curve characteristics that they were recognizable at all protein levels. The authors concluded that the curve's greatest value, provided that protein content was known, was to characterize the type to which the flour belongs. The curves established qualitative differences between wheats that may or may not be equal in strength and thus indicated probable baking performance.

Swanson (1941) investigated the effects of protein content and absorption on mixograph curves. He concluded that the major factor affecting curve pattern was variety, and that within a variety, protein content was the most potent factor. Given correct absorption the height of a curve will increase with increasing protein. An absorption 2% above or below optimum had little effect on varietal pattern but did affect peak height.

The first systematic and statistical study relating mixogram characteristics to baking results was conducted by Johnson et al (1943). Peak height, width and weakening angle were positively correlated with protein content and loaf volume. The authors felt that mixogram characteristics tended to reflect baking strength of a flour, mainly because of the high correlation between loaf volume and protein, and the correlation between protein and mixogram characteristics. The angle between ascending and descending portions of the curve tended to decrease with increasing protein but not in a linear fashion. The relationship of mixing time and of development angle with protein content was curvilinear.

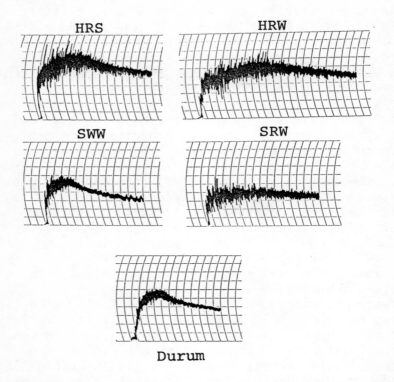

Figure 9. Effect of wheat class on mixogram characteristics when absorption is optimized by using farinograph absorption.

Development angle and mixing time were moderately and negatively correlated. Area was not significantly correlated with loaf volume. Johnson et al concluded that the most important use of the mixograph was to furnish information which supplements baking data, for example, mixing requirements, mixing tolerance and varietal pattern.

Morris et al (1944) investigated the use of the mixograph in evaluating the quality of flour from soft wheat varieties. A single figure score was obtained by measuring the area under the curve

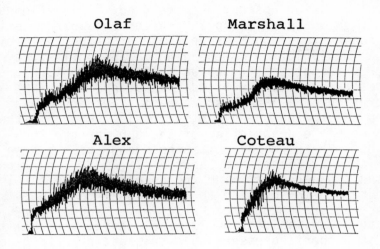

Figure 10. Effect of variety on mixogram characteristics of HRS wheats comparably grown (67% absorption).

obtained from a mixing time of 7 minutes. The area was suggested as a general expression of strength due to gluten quality and quantity. In a test of 29 varieties the area could be used to rank the varieties in an order consistent with the relative quality of the varieties and area was correlated with the results of other quality tests. However, since the varietal quality characteristics as expressed by the mixograms were affected greatly by environment, comparative evaluation of soft wheat varieties could be made only on samples grown under uniform conditions.

Although the mixograph is extensively used, its value is limited because of several factors. These factors relate to the fact that the mixograph is not a standardized instrument. Under uniform conditions in any one laboratory, a usable level of replication can be obtained. However, in interlaboratory collaborative work the variability is high. A portion of the problem is that although there is an AACC method for use of the mixograph, the method contains only a general description of the instrument's use. For example, varied absorptions are used. Bread flours are often analyzed at an absorption considered optimum for a bread dough as determined subjectively by manual manipulation of the dough or objectively by other physical tests such as protein content. Some operators have used a fixed absorption; e.g., 66% absorption at a 14% moisture basis. Others use a fixed absorption on an "as is" basis. Probably the most complete description of the appropriate techniques to be used for obtaining mixograms has been published by Finney and Shogren (1972) as a portion of their paper describing the development of the 10 g mixograph. In addition to operational techniques other factors such as operator, mixer speed, temperature, spring tension setting, spring variability, and atmospheric pressure have substantial effects on the curves obtained and must be considered.

In work conducted at the USDA/ARS Wheat Quality Lab the effect of absorption on peak time, peak height and area was studied. The effects over the range of 60-80% absorption are shown in Table IV for HRS wheat flour. Absorption was positively correlated with peak time and area ($r = 0.935$ and 0.860, respectively) and negatively correlated with peak height ($r = 0.991$).

Harris (1943) conducted a comparative study of the estimation of dough development, dough stability and curve height of mixograms by three operators on 10 hard red spring wheat flours adjusted to a uniform protein content of 12%. Significant differences between operators were observed for dough development and curve height but not for stability. However, differences in curve properties among samples were much greater than those due to operators except for curve height. Curve heights would have been different, if original flours of varied protein contents had been used.

TABLE IV
Effect of Absorption on Some
Mixogram Characteristics

Absorption (%)	Mixograph Parameter		
	Peak Time (mm)	Peak Height (mm)	Area (in^2)
60	72	85	8.59
65	94	81	9.53
70	125	70	9.75
75	169	65	11.44
80	291	57	15.25

Baig and Hoseney (1977) studied the effect of mixer speed, dough temperature and absorption on mixograms. Increasing mixing speed decreased mixing time with the effect being more pronounced at higher temperatures. Higher mixer speed appeared to increase curve height, but this probably occurred due to the mixer pins striking the dough more often rather than from increased dough resistance.

The effects of temperature on mixograms have been reported (Harris et al 1944, Harris and Sibbitt 1945, Heizer et al 1951, Baig and Hoseney 1977). In general, increasing temperature decreases curve height, width and area. Shown in Table V is the effect of temperature on peak height, peak time and area. Peak time is positively linearly correlated with temperature ($r = 0.929$) and peak height and area are negatively correlated with temperature ($r = 0.998$ and 0.980, respectively).

Heizer et al (1951) investigated the effect of spring variability on mixograms. When eight mixograph springs were compared to a standard calibration spring, spring length and spring extension under load varied. They concluded that spring length could be adjusted for, but that springs differing in extension properties were comparable only

TABLE V
Effect of Temperature on Some Mixogram Characteristics

Temperature (°C)	Mixograph Parameter		
	Peak Time (mm)	Peak Height (mm)	Area (in^2)
20	94	96	11.43
25	97	81	9.56
30	113	64	8.69
35	167	44	7.75
40	251	30	7.06

at their point of standardization. Shuey and Gilles (1966) investigated the effect of different spring settings. At settings of 6 and 8 the torque required to displace the pen increased in a curvilinear fashion and at settings of 10 and 12 in a linear fashion.

Lorenz (1974) studied the effect of varying atmospheric pressure on mixogram characteristics. Peak height, peak time, and the area under the curve increased as elevation increased. The weakening angle was not affected.

The mixograph has been used for various purposes since its initial appearance. It can be used to determine optimum absorption and mixing time prior to baking. At the USDA/ARS Wheat Quality Lab all spring wheat mixograms are obtained on 30 g flour at a constant 66% absorption. Based on the peak height, a value for baking absorption can be obtained from a chart which was established experimentally. The baking mixing time is based on a number calculated as the peak time plus or minus a correction factor for peak height. The number obtained is linearly related to mixing time and is also used to assign a mixogram score to the flour. This system has been in use in the ARS lab for more than twenty years. North Dakota

State University's Department of Cereal Science and Food Technology uses 25 g flour samples at 60% absorption to obtain spring wheat mixograms. The mixograms are compared to previously established standard curves for assignment of scores, and strong, weak or abnormal curves are noted. For evaluation of durum semolina, both organizations utilize a constant absorption on an as-is basis. The resulting mixograms are compared to a set of standard curves to which numerical values have been assigned.

One use of the mixograph, which has appeared often in the literature, is its use to evaluate wheat quality based on the curves obtained from ground whole wheat (Johnson and Swanson 1942, Lamb 1944, Sibbit and Harris 1945, Bruinsma et al 1978). The purpose of using wheat meal instead of flour is to avoid the time and expense involved in preparing a refined flour as well as allowing evaluation of small wheat samples, such as early generation breeders' samples. In work conducted at the USDA/ARS Wheat Quality Lab, a high correlation was observed between the areas and peak times observed for spring wheat flours and the areas and peak times observed for whole wheat meals ($r = 0.916$ and 0.949, respectively). Also, the correlations between mixograph score obtained as described earlier, and the wheat meal peak times and areas are fairly high ($r = 0.925$ and 0.900, respectively).

In conclusion, both the mixograph and farinograph provide information concerning the quality of wheat in relation to mixing characteristics. The information derived, however, must be supplemented with other quality tests including the baking test in order to establish the overall quality of a wheat sample.

LITERATURE CITED

Baig, M.M. and Hoseney, R.C. 1977. Effects of mixer speed, dough temperature, and water absorption on flour-water mixograms. Cereal Chem. 54:605.

Bruinsma, B.L., Anderson, P.D. and Rubenthaler, G.L. 1978. Rapid method to determine quality of wheat with the mixograph. Cereal Chem. 55:732.

Bushuk, W. 1966. Distribution of water in dough and bread. Bakers Digest 40(5):38.

D'Appolonia, B.L. and Kunerth, W.H. editors. 1984. The Farinograph Handbook, American Associaiton of Cereal Chemists, St. Paul, MN.

Finney, K.F. and Shogren, M.D. 1972. A ten-gram mixograph for determining and predicting functional properties of wheat flours. Bakers Digest 46(2):32.

Harris, R.J. 1943. The effect of different operators on the evaluation of mixograms. Cereal Chem. 20:739.

Harris, R.H. and Sibbitt, L.D. 1945. Comparative effects of absorption and temperature in mixogram patterns of different wheat varieties. Bakers Digest, 19(1):17.

Harris, R.H., Sibbitt, L.D. and Scott, G.M. 1944. The effect of temperature differences on some mixogram properties of hard red spring wheat flours. Cereal Chem. 21:374.

Heizer, H.K., Moser, L., Bode, C.E., Yamazaki, W.T. and Kissell, L.T. 1951. Studies concerning mixograph standardization. American Association of Cereal Chemists Transactions 9:16.

Johnson, J.A. and Swanson, C.O. 1942. The testing of wheat quality by recording dough mixer curves obtained from sifted wheat meals. Cereal Chem. 19:216.

Johnson, J.A., Swanson, C.O. and Bayfield, E.G. 1943. The correlation of mixograms with baking results. Cereal Chem. 20:625.

Lamb, C.A. 1944. Sifted wheat meal mixograms for selecting soft wheat varieties. Cereal Chem. 21:57.

Landis, Q. and Freilich, J. 1935. Studies on test dough mixer calibrations. Cereal Chem. 12:665.

Larmour, R.K., Working, E.B. and Ofelt, C.W. 1939. Quality tests on hard red winter wheats. Cereal Chem. 16:733.

Lorenz, K. 1974. Mixogram characteristics as affected by varying atmospheric pressures. Cereal Sci Today 19:322.

Morris, C.E., Bode, C.E. and Heizer, H.K. 1944. The use of the mixogram in evaluating quality in soft wheat varieties. Cereal Chem. 21:49.

Sibbitt, L.D. and Harris, R.H. 1945. Comparisons between some properties of mixograms from flour and unsifted whole meal. Cereal Chem. 22:531.

Sibbitt, L.D., Harris, R.H. and Conlon, T.J. 1953. Some relations between farinogram and mixogram dimensions and baking quality. Bakers Digest 27(4):26.

Shuey, W.C. 1974. Practical instruments for rheological measurements on wheat products. Cereal Chem. 52:42r.

Shuey, W.C. and Gilles, K.A. 1966. Effect of spring settings and absorption on mixograms for measuring dough characteristics. Cereal Chem. 43:94.

Swanson, C.O. 1941. Factors which influence the physical properties of dough. III. Effect of protein content and absorption on the pattern of curves made on the recording dough mixer. Cereal Chem. 18:615.

Swanson, C.O. and Johnson, J.A. 1943. Description of mixograms. Cereal Chem. 20:38.

Swanson, C.O. and Working, E.B. 1933. Testing quality of flour by the recording dough mixer. Cereal Chem. 10:1.

CONSTANT WATER CONTENT vs. CONSTANT CONSISTENCY TECHNIQUES IN ALVEOGRAPHY OF SOFT WHEAT FLOURS

V.F. Rasper and K.M. Hardy
Department of Food Science
University of Guelph
Guelph, Ontario, N1G 2W1
Canada

G.R. Fulcher
Agriculture Canada
Ottawa Research Station
Ottawa, Ontario, K1A 0C6
Canada

INTRODUCTION

The alveography is one of the rheological techniques designed for a routine testing of wheat flours. A dough prepared from tested flour under standard conditions of water addition and mixing is sheeted and cut into a circular test piece which, after a period of resting, is subjected to biaxial extension by inflating it into the shape of a bubble until it ruptures. The pressure in the bubble is measured by a manometer and recorded on a chart as a function of time. In Fig. 1, the most recent model of the Chopin Alveograph MA 82 is shown. This model has a built-in air pump which provides air for stretching the dough bubble at an easily calibrated flow rate. With the older models, still in common use, the dough bubble is inflated by air displaced from a reservoir by water which is driven through a constriction under a slowly changing pressure head. Shown in Fig. 1 are

Fig. 1. Chopin Alveograph MA 82
(Courtesy of Tripette and Renaund S.A., 1985)

also utensils necessary for the preparation of the dough test pieces such as the sheeting assembly and the cutter.

The biaxial extension produced during the dough bubble inflation has often been pointed out as a distinct advantage of the alveographic test over stretchability tests in which dough samples are subjected to deformations in simple tension. From a physical viewpoint, biaxial extension is well linked with the gas cell expansion in the rising dough. Thus, the alveograph test may produce data relevant for predicting the dough behaviour during fermentation and the early stages of baking and may, therefore, serve as a valid tool in assessing the overall baking quality of the tested flour. A representative pressure/time record,

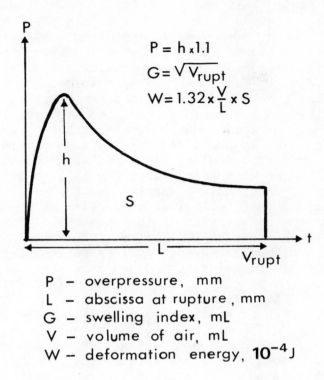

Fig. 2. Representative alveogram showing some commonly measured indices

or alveogram, and the indices considered of practical siginificance are presented in Fig. 2. The interpretation of these indices in terms of flour quality is based on some empirically established relationships between the shape of the alveograms and the actual dough performance during the baking process. Needless to say, this interpretation still remains, to a great extent, a matter of an individualistic approach. It is reasonable to suppose that whereas the fundamental principles will always be the same, the practical relevance of the different indices will greatly depend on the nature of both the technological process and the final product for which the tested flour is intended.

Scott Blair and Potel (1937) found the overpressure $\underline{P}$ related to the dough viscosity and considered it a measure of the water absorbing capacity of the flour. Bloksma (1957a), when calculating the shape of the alveograms for some rheological model substances, did not find the relationship between $\underline{P}$ and viscosity valid for Maxwell bodies. His calculations for the model substances revealed a decreasing trend in the height of the alveograms with increasing relaxation time in spite of the viscosity remaining constant. The researcher agreed with the earlier conclusions of Hlynka and Barth (1955) that the maximum height in a normal alveogram need not to be a consequence of the rheological properties of the tested dough, but may depend on the geometry of the experiment. In a search for a more useful rheological index, he suggested the use of a ratio between the height of the curve at a dough bubble volume of 100 mL and the height at maxmimum. This ratio was expected to reflect the relaxation time of the tested dough. The suggestion was made with reference to the work of Halton and Scott Blair (1936) who state that a good bread dough had, among other characteristics, a sufficiently long relaxation time. However, a relatively poor correlation was found when the P_{100}/P_{max} ratio was plotted against loaf volumes of a great number of test breads (Bloksma 1957b). This failure in establishing the expected relationship was considered a result of a simultaneous involvement of both a desirable long relaxation time and of a harmful structural viscosity (Bloksma 1958).

Aitken et al. (1944a, 1944b) related the maximum height of the alveograms to empirical terms such as stiffness, shortness and tightness. Amos (1949), from a study on rheological methods in milling and baking, concluded that the overpressure $\underline{P}$ could be used as an index of dough stability. The recently accepted ISO Alveograph Method (ISO 1983) defines overpressure $\underline{P}$ as an indicator of the resistance of dough to deformation.

There seems to be less ambiguity in the interpretation of the length of the curve, the average

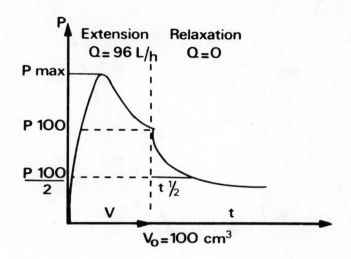

Fig. 3. Relaxation time test with the Chopin Relaxometer (Launay 1979)

abscissa $\underline{L}$, which is generally taken as a measure of dough extensibility. The proportional number obtained by dividing the overpressure $\underline{P}$ with the average abscissa $\underline{L}$ was also suggested by several investigators as an useful means of predicting flour quality (Maes and Pirotte 1957, Chopin 1962). Much weight was attached to the area under the curve $\underline{S}$ from which the deformation energy $\underline{W}$, representing the energy necessary to inflate the dough until it ruptures, can be derived (Chopin 1927). This energy is often taken as an indicator of flour strength. The significance of the swelling index $\underline{G}$ was not so easy to establish. According to Scott Blair and Potel (1937), this index is primarily related to a complex function of viscosity and modulus. In terms of empirical dough rheology, it may be considered approximately dependent on the product of properties usually described as springiness and shortness.

With the Chopin Alveograph MA 82, an integrating calculator can be employed for a digital readout or,

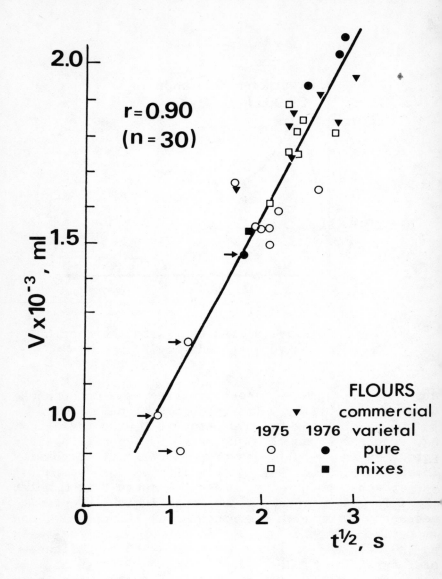

Figure 4. Relationship between relaxation time determined by Chopin relaxometer and loaf volume values (pain français) for 30 samples of French flours (Launay 1979).

→ Flours unsuitable for bread baking.

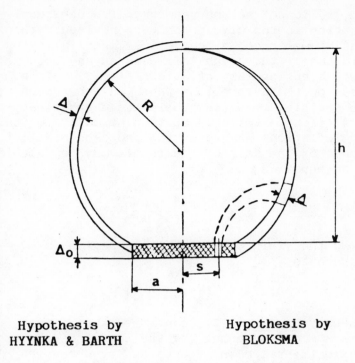

Hypothesis by	Hypothesis by
HYYNKA & BARTH	BLOKSMA

Figure 5. Geometry of the dough bubble inflation
(Launay and Buré 1970)

if linked with a printer, a printout of all the above
mentioned indices. The model can also be used in
combination with a relaxometer for relaxation time
testing, the principle of which is graphically demonstrated in Figure 3. As reported by Launay (1979),
relaxation times determined by this procedure correlated very closely with loaf volumes of French breads
(Fig. 4).

Although the alveographic technique was orginal-
ly designed as an empirical tool for routine testing,
several studies were carried out with the objectives
of interpreting the results in terms of fundamental
rheology using the standard physical units. Such interpretation can only be made possible if the mode
and rate of deformation of the tested material are
well defined. Early attempts by Hlynka and Barth

(1955a, b) to define the geometry of dough bubble deformation in the alveograph were revised by Bloksma (1957a) who proposed a theoretical model allowing the calculation of the dough thickness at any point of the bubble during its inflation (Fig. 5). This model is based on the assumptions that (a) the dough is incompressible, (b) the bubble has a spherical shape and (c) each particle is shifted normally to itself during inflation. Based on the proposed model, the thickness of the wall of the bubble may be calculated using an equation

$$\Delta = \Delta_o \left\{ \frac{a^4 + s^2 h^2}{a^2(a^2 + h^2)} \right\}^2 \qquad [1]$$

which in the vicinity of the polar region ($s^2 \ll a^4/h^2$) is reduced to

$$\Delta = \Delta_o (1 + h^2/a^2)^{-2} \qquad [2]$$

Launay et al. (1977) confirmed the validity of Bloksma's model by comparing the calculated values with experimental results obtained by chronographic recording of bubble expansion. They found a good agreement between the compared data, especially with small rates of expansion and small volumes. With a defined deformation of the dough bubble, Launay and Buré (1977) were able to evaluate the stress (σ), strain (ε) and strain rate ($\dot{\varepsilon}$) during the inflation. From their data they concluded that in equal biaxial extension, wheat flour dough follows a power law, at least with $\dot{\varepsilon}$ between 10^{-2} and 1 s^{-1}. The power law exponent did not appear to change very much from flour to flour, but there was a definite trend towards smaller values for stronger flours. The researchers were also able to confirm Schofield and Scott Blair's (1933) observations of the ε-hardening effect which, presumably, was responsible for Bloksma's (1957a) unsuccessful attempt to calculate the shape of alveograms using simple rheological models. The phenomenon of ε-hardening became very pronounced beyond $\varepsilon = 1$ (170% extension).

Some critical comments pertaining to the fundamental alveographic studies were raised by Hibberd and Parker (1974) who investigated the rate of growth of dough bubble using the classical model of the instrument with water displacement system. By means of a pressure transducer coupled with the water manometer, they found the effects of the displacement of water in the manometer and the compression of the air trapped between the water from the alveograph burette and the sheet of dough significant enough not to be neglected, as it was done by previous researchers. They, however, emphasized that such detailed considerations were only relevant in discussions concerning the fundamental apsects of dough behaviour and had no measurable impact on results of routine quality testing.

The Chopin Alveograph was first used as early as the twenties (Chopin 1927) but its application for wheat and wheat flour quality testing was for many years confined to a few European countries, mainly to its country of origin. Nowadays, the instrument is used in cereal laboratories all over the world. In North America, an increasing use is made of alveography as a criterion of soft wheat and soft wheat flour quality in replacement for the hitherto very common MacMichael viscosity test. With all this growing interest in alveographic technique, some of its aspects, which were the subject of numerous disputes among the dough rheologists several decades ago, have once again been brought into the focus of our attention. As in the previous years, concern has been expressed about using doughs of a constant water content without allowing for differences in the hydration capacity of the tested flours. A constant mixing time and a relatively short resting period have also been a frequent target for criticism. From all the earlier reported efforts to elucidate the practical relevance of the disputed factors (Tchetveroukhine 1947, 1948; Marcelle 1955; Maes and Pirotte 1957, Nuret et al. 1970, Khattak et al. 1974, Weipert 1981), it became clearly evident that their impact on the final outcome of the test can only be judged relative to the type of the tested flour and its rheological contribution to the system in which it will be involved. It should be noted that most of the so-far published work concerning the disputed aspects dealt with stronger flours intended for fermented doughs. Conclusions from such studies may not be necessarily fully applicable to soft wheat flours and their functional behaviour in systems like cake batters and cookie doughs.

EXPERIMENTAL

In the current study on the use of alveography in the quality assessment of soft wheats, the main emphasis was put on comparing data obtained by the standard procedure, i.e. using doughs of constant water content, with results on doughs mixed to an identical consistency under the conditions of a variable water/flour ratio. The standard alveograph procedure was used as described in the recently approved AACC method 54-30. In preparing doughs of a variable water content, the addition of water was adjusted so that the doughs reached a consistency of 500 B.U. upon maximum development when mixed in a 300-g farinograph mixing bowl following the dough preparation procedure of the AACC extensigraph method 54-10 (AACC 1983). The doughs were transferred into the alveograph mixer for extrusion after which they were handled in the standard manner.

In the first set of experiments, the two procedures were compared on flours milled from 42 soft wheat varieties grown at three locations in Ontario in 1984, and 7 varieties grown in 1983. All flours were milled on a Buhler mill MLU 202, supplemented with a bran finisher, to an average yield of 72.2%. Apart from alveograph testing, the flours were subjected to tests presently used as part of the quality assessment methodology for Ontario grown soft wheats. The testing involved the AACC procedures for the determination of protein, MacMichael viscosity, starch damage and cookie diameter (AACC 1983). Alkaline water retention capacity was determined by the Yamazaki's method (Yamazaki 1953).

Table 1 gives the mean values and ranges of the individual quality indicators for all tested flours.

TABLE 1
Quality Testing of Varietal Soft Wheat Flours
Grown in Ontario, 1984, 1983 (n=49)

	Range	Mean	S.D.[a]
Protein (%)	7.2 - 9.1	8.2	0.48
MacMichael Viscosity (°M)	28 - 76	45.8	11.5
AWRC[b] (%)	63.2 - 71.7	64.9	9.24
Starch Damage (%)	1.50 - 2.76	1.97	0.30
Farinograph Absorption[c] (%)	47.5 - 49.5	48.6	0.53
Cookie Diameter (cm)	8.6 - 9.2	8.9	0.12

[a] Standard deviation.
[b] Alkaline water retention capacity (Yamazaki 1953).
[c] Determined in the presence of 2% NaCl on flour weight basis.

Alveogram data obtained under the conditions of the two dough preparation procedures are summarized in Table 2. Doughs of constant consistency gave $\underline{P}$ and $\underline{W}$ values consistently higher than doughs prepared in the standard manner. They were also characterized by a reduced extensibility $\underline{L}$ and slightly lower values of swelling index $\underline{G}$. These differences in the alveogram indices may be, in the first place, considered as a consequence of a reduced water content. Doughs prepared in the standard manner had a water content of 43.0%, while water content in doughs characterized by an identical consistency ranged from 41.7 to 42.5% with an average value of 42.1%. This difference in water content, however, did not seem to be high

TABLE 2
Alveogram Data for Varietal Soft Wheat Flours
Grown in Ontario, 1984, 1983 (n=49)

		Range	Mean	S.D.[b]
Overpressure, $\underline{P}$ (mm)	CW[a]	16.1 - 27.3	20.0	2.79
	CC[a]	29.3 - 37.4	39.5	4.95
Average Abscissa, $\underline{L}$ (mm)	CW	86 - 201	132.3	34.3
	CC	65 - 163	105.7	22.8
P/L	CW	0.11 - 0.30	0.16	0.04
	CC	0.18 - 0.57	0.34	0.09
Swelling Index, $\underline{G}$ (mL)	CW	17.0 - 28.8	24.0	3.17
	CC	18.0 - 28.5	22.7	2.63
Deformation Energy, $\underline{W}$ (10^4 xJ)	CW	21.0 - 71.7	43.9	13.06
	CC	40.4 - 101.4	70.1	14.26
Water in Dough (% of Flour Dry Solids)	CW	---	43.0	--
	CC	41.7 - 42.5	42.1	0.21

[a] CW = constant water content; CC = constant dough consistency.
[b] Standard deviation.

enough to be entirely responsible for the recorded differences in the alveogram data. It may be reasonable to suggest that the longer mixing time of the standard procedure was also a contributing factor. While doughs prepared by this procedure received a total of 6-min mixing time, mixing time for doughs of constant consistency did not exceed a total of 3 minutes. Longer mixing is known to result in lower resistance to extension and enhanced extensibility (Bailey and LeVesconte 1924, Khattak et al. 1974).

In spite of these differences in the alveogram

TABLE 3
Correlation Coefficients for Alveogram Data of
Varietal Flours Tested Under Two Different
Conditions of Dough Preparation

	W_{cc}	P_{cc}	L_{cc}	P/L_{cc}	G_{cc}
W_{cw}	0.86**				
P_{cw}		0.64**			
L_{cw}			0.42*		
P/L_{cw}				0.23	
G_{cw}					0.44*

*Significantly different at $P < 0.05$, **$P < 0.01$.

Subscripts cw and cc indicate the constant water content and constant dough consistency procedure, respectively.

indices, there was a strong correlation found for values obtained under the two conditions of dough preparation (Table 3). The only exception was the P/L ratio. The strongest correlation was associated with W values ($r=0.86$, $P<0.01$). A graphical plot of these values is presented in Fig. 6.

Table 4 shows how the individual alveogram indices correlated with the other determined qualities of the tested flours. With the exception of overpressure P determined under the conditions of constant dough consistency, they were all closely related to the flour protein content. Similar relationships were reported earlier by Aitken et al. (1944b), whose findings, however, were not supported by results of Khattak et al. (1974) on HRS flours. High correlation was also found between the alveogram data and MacMichael viscosity, except of P/L. The

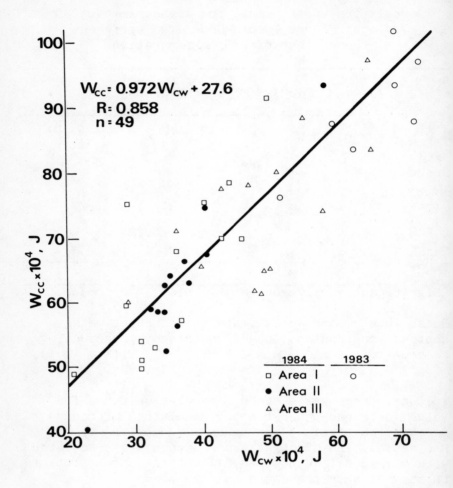

Fig. 6. Relationship between the $\overline{W}$ values determined for soft wheat flours under two different conditions of dough preparation.

TABLE 4
Correlation Coefficients for Alveogram Data and Other Quality Criteria for Varietal Soft Wheat Flours (n = 49)

	Protein	MacMichael Viscosity	Farinograph Absorption[a]	AWRC[b]	Cookie Diameter
P_{cw}	0.40**	0.70**	0.03	0.32*	0.17
P_{cc}	0.10	0.59**	0.06	-0.09	-0.07
L_{cw}	0.53**	0.54**	-0.01	0.20	-0.36*
L_{cc}	0.54**	0.37*	0.37*	0.47**	0.22
P/L_{cw}	-0.33*	-0.14	0.25	0.05	0.24
P/L_{cc}	-0.46**	-0.27	-0.40	-0.39*	0.12
G_{cw}	0.56**	0.51**	-0.01	0.18	-0.37*
G_{cc}	0.52**	0.41**	0.31*	0.46**	0.13
W_{cw}	0.78**	0.87**	0.46**	0.28	-0.23
W_{cc}	0.60**	0.82**	0.36*	0.26	-0.16
Protein	--	0.89**	0.34*	0.51**	-0.21
MacMichael Viscosity	--	--	0.36*	0.39**	-0.22
Farinograph Absorption[a]	--	--	--	0.18	0.10
AWRC[b]	--	--	--	--	-0.28

[a] Determined in the presence of 2% NaCl on flour basis.
[b] Alkaline water retention capacity (Yamazaki 1953). Subscripts cw and cc indicate the constant water content and constant dough consistency procedure, respectively.
*Significantly different at $P < 0.05$, **$P < 0.01$.

dough preparation procedure did not seem to have any effect on any of these correlations, though in most cases, they appeared somewhat stronger when data recorded under the conditions of constant water content were involved.

The relationships between the measured alveogram data and quality attributes other than those mentioned above appeared to be less ambiguous. In disagreement with findings of Weipert (1981), no significant correlation was found between the height of the alveograms and farinograph absorption. The plot of farinograph absorption vs. deformation energy yielded a correlation coefficient of 0.46 ($P<0.01$) and 0.36 ($P<0.05$) for the constant water content and constant consistency procedure, respectively. Indices based on the length of the alveogram, i.e. average abscissa $\underline{L}$ and swelling index $\underline{G}$ as well as $\underline{P/L}$, correlated ($P<0.05$) with farinograph absorption only when read from curves obtained by testing doughs of constant consistency. The same was observed when these indices were compared with the AWRC values. The correlations, however, appeared stronger ($P<0.01$) than in the case of farinograph absorption. As for the cookie diameter, the only significant correlation was found with $\underline{L}$ and $\underline{G}$ values for doughs of constant water content. Neither correlation, however, was strong enough to be a good predictor of cookie diameter.

This poor relationship between the cookie diameter and other qualities of the tested flours was not surprising. With a narrow spread in the diameter values (8.6-9.2 cm) and a relatively low sensitivity of the test itself (Abboud et al. 1985), no high correlations were expected. In order to broaden the variability of the tested material, some of its properties were modified under laboratory controlled conditions. A series of flour samples were prepared from SWW wheat after it had been soaked in water for 4 h at 24°C and let germinate for 6 to 30 h at the same temperature. The amylograph diastatic activity test (AACC 1983) was used to assay the straight-grade flours milled from the germinated grain on Buhler mill MLU 202, for the extent of their modification.

TABLE 5

Cookie Spread, MacMichael Viscosity and Alveogram Data for Flours with Varying Diastatic Activity

Diastatic Activity[a] B.U.	Cookie Spread Factor	MacMichael Viscosity °M		Alveogram Data						Water in Dough[c] %
				P mm	L mm	P/L	G mL	W 10^4J		
550 (0)	92.3	31	CW[a] CC[b]	23.6 42.9	116 105	0.20 0.41	23.5 22.5	64.2 107.8		43.0 42.0
520 (6)	93.4	31	CW CC	21.0 40.8	155 108	0.13 0.38	27.5 23.5	62.7 104.6		43.0 42.0
510 (12)	92.0	30	CW CC	22.3 41.1	149 112	0.15 0.37	27.0 33.5	63.4 104.0		43.0 42.0
200 (18)	91.2	31	CW CC	19.5 35.2	141 115	0.14 0.32	26.0 24.0	50.8 97.1		43.0 42.0
90 (24)	93.4	31	CW CC	19.3 33.5	139 134	0.14 0.25	26.0 25.5	46.4 87.8		43.0 41.9
50 (30)	93.4	31	CW CC	18.4 31.9	164 122	0.11 0.26	28.5 24.5	45.6 81.6		43.0 41.8

[a] Determined by Brabender ViscoAmylograph; figures in parenthesis indicate time of germination.
[b] CW = constant water content; CC = constant dough consistency.
[c] NaCl not included.

The original activity of 500 B.U. of the nongerminated sample was reduced to 30 B.U. after 30 h germination (Table 5). Alveograms of the tested samples responded to this change by a reduced height and increased length of the curve. The increase in the latter, however, was not high enough to compensate for the reduction in height and, as a result, there was a steady decrease in the area under the curve and, consequently, deformation energy $\overline{W}$ as the germination process progressed. All these changes in the alveogram data with the depth of germination followed the same pattern regardless of the dough preparation procedure.

Qualitative changes in flours milled from the germinated grain, which were so markedly reflected in the results of alveograph testing, remained practically unrevealed by either MacMichael viscosity test or cookie spread test.

Relationships different from those which emerged from experiments with flours modified by germination were established from results on flours characterized by a variable degree of starch damage. Flour samples with a varying percentage of damaged starch were prepared by subjecting a commercial cookie flour (3.9% damaged starch) to treatment in a SWECO Vibroenergy Mill FM-1 (Sweco Inc., Los Angeles, CA). Starch damage content in the treated samples ranged from 8.5 to 18.0%. Both MacMichael viscosity and cookie spread factor responded very strongly to this treatment (Table 6). An equally significant response was seen in the height and length of the alveograms. With increasing starch damage, there was an increase in the overpressure $\overline{P}$ accompanied by a decrease in the average abscissa $\overline{L}$ and swelling index $\overline{G}$. It can be noticed that counterbalancing the increased hydration capacity of damaged flours by raising the amount of added water to a level producing dough of 500 B.U. consistency, did not prevent the increase in the height of the curves. Unlike experiments with varietal flours, $\overline{P}$ values of doughs prepared by both procedures were closely related to farinograph absorptions of the tested flours which ranged from 50.7 to 63.6% (in the presence of 2% NaCl on flour basis).

TABLE 6

Cookie Spread, MacMichael Viscosity and Alveogram Data for Flours with Varying Degree of Starch Damage

Starch Damage %	Cookie Spread Factor	MacMichael Viscosity °M		Alveogram Data					Water in Dough[b] %
				P mm	L mm	P/L	G mL	W 10^4J	
3.9	92.6	35	CW[a]	27.8	123	0.23	24.5	67.9	43.0
			CC[a]	34.8	95	0.38	21.6	86.3	42.9
8.5	66.5	38	CW	53.6	64	0.85	17.6	104.8	43.0
			CC	46.4	86	0.55	20.5	112.4	44.4
13.6	57.9	41	CW	87.5	49	1.82	15.5	146.5	43.0
			CC	56.9	56	1.01	16.5	111.4	46.0
18.0	52.8	44	CW	125.8	31	4.00	12.4	160.8	43.0
			CC	60.2	50	1.22	15.1	112.3	47.4

[a]CW = constant water content; CC = constant dough consistency.
[b]NaCl not included.

The closeness of this association was summarized by correlation coefficients of r = 0.998 r = 0.973 for the constant water content and constant dough consistency procedures, respectively. The increase in $\underline{P}$, however, was not so steep with doughs having the adjusted water content as with those prepared in the standard manner. Linear regression for the latter was characterized by a slope of 7.76, whereas with doughs of constant consistency a slope value of 1.99 was calculated.

Unlike $\underline{P}$, $\underline{L}$ or $\underline{G}$, the response of the deformation energy $\underline{W}$ to the increasing severity of starch damage was found to be greatly influenced by the dough preparation procedure. Under the conditions of constant water content, there was an increasing trend as more severe damage was inflicted upon the starch granules. With doughs of constant consistency, $\underline{W}$ increased after the first interval of treatment in the vibratory mill (8.5% damaged starch), but remained practically unchanged for the rest of the experiment.

SUMMARY

Reported results indicated a satisfactory sensitivity of the Chopin Alveograph in responding to the variability in the quality of the tested soft wheat flours. Due to the closeness in the quantity of water which was added to the flours when following the standard procedure, and that required for bringing the tested dough to a consistency of 500 B.U., the constant consistency procedure produced results which rated the flours in practically the same order of their quality as the standard test. The only exception was the case when the flour hydration capacity was altered to an extent that it exceeded values usually associated with flours possessing properties within the normal range of quality specifications.

ACKNOWLEDGEMENT

Experimental work reported in this paper was part of a project jointly supported by Agriculture

Canada and Ontario Ministry of Food and Agriculture. The authors wish to acknowledge Mrs. Lourdes Pico and Miss Ruth Spark for their technical assistance and Mr. M. Dubois, Tripette & Renaud S.A. (Paris) for valuable comments on the use of the instrument.

LITERATURE CITED

AMERICAN ASSOCIATION OF CEREAL CHEMISTS, 1983. Approved Methods of the AACC. Methods 10-50D, 22-10, 46-12, 54-10, 54-21, 56-80, 76-30A. The Association, St. Paul., MN.

ABBOUD, A.M., RUBENTHALER, G.L., and HOSENEY, R.C. 1985. Effect of fat and sugar in sugar-snap cookies and evaluation of tests to measure cookie flour quality. Cereal Chem. 62: 125-129.

AITKEN, T.R., FISHER, M.H., and ANDERSON, J.A. 1944a. Reproducibility studies of some effects of technique on extensograms and alveograms. Cereal Chem. 21: 489-498.

AITKEN, T.R., FISHER, M.H., and ANDERSON, J.A. 1944b. Effect of protein content and grade on farinograms, extensograms and alveograms. Cereal Chem. 21: 465-488.

BAILEY, C.H., and LEVESCONTE, A.M. 1924. Physical tests of flour quality with Chopin Extensimeter. Cereal Chem. 1: 38-63.

BENNETT, R., and CAPPOCK, J.B.M. 1952. Measuring the physical characteristics of flour. A method of using the Chopin Alveograph to detect the effect of flour improvers. J. Sci. Fd. Agr. 3: 297.

BLOKSMA, A.H. 1957a. A calculation of the shape of the alveograms of some rheological model substances. Cereal Chem. 34: 126-136.

BLOKSMA, A.H. 1957b. L'emploi de l'alvéographe Chopin pour la determination de la valeur boulangère des farines. Inds. aliment. et agr. (Paris) 74: 653-657.

BLOKSMA, A.H. 1958. A calculation of the shape of the alveograms of materials showing structural viscosity. Cereal Chem. 35: 323-330.

CHOPIN, M. 1927. Determination of baking value of wheat by means of specfic energy of deformation of dough. Ceral Chem. 4: 1-13.

CHOPIN, M. 1962. Sur l'utilization du rapport P/L dans l'essai des farines avec l'alvéographe. Bull. de l'Ecole Française de Meunerie No. 189, 139-141.

HALTON, P., and SCOTT BLAIR, G.W. 1936. A study of some physical properties of flour doughs in relation to their breadmaking qualities. J. Phys. Chem. 40: 561-580.

HIBBERD, G.E., and PARKER, N.S. 1974. The rate of growth of dough bubbles on the Chopin Alveograph. Lebensmitt.-Wiss. u. Technol. 7: 318-321.

HLYNKA, I., and BARTH, F.W. 1955a. Chopin alveograph studies. I. Dough resistance at constant sample deformation. Cereal Chem. 32: 463-471.

HLYNKA, I., and BARTH, R.W. 1955b. Chopin alveograph studies. II. Structural relaxation in dough. Cereal Chem. 32: 472-480.

INTERNATIONAL ORGANIZATION FOR STANDARDIZATION, 1983. Wheat flour - Physical characteristics of doughs Part IV: Determination of rheological properties using an alveograph. International Standard ISO 55 30/4 - 1983(E).

KHATTAK, S., D'APPOLONIA, B.L., and BANASIK, O.J. 1977. Use of alveograph for quality evaluation of hard red spring wheat. Cereal Chem. 51: 355-363.

LAUNAY, B. 1979. Propriétés rhéologiques des pâtes de farine: Quelques progrès recents. Inds. aliment. et agr. 96: 617-623.

LAUNAY, B., and BURÉ, J. 1970. Alvéographe Chopin et propriétés rhéologiques des pâtes. Lebensm.-Wiss. u. Technol. 3: 57-62.

LAUNAY, B., and BURÉ, J. 1977. Use of the Chopin alveographe as a rheological tool. II. Dough properties in biaxial extension. Cereal Chem. 54: 1152-1158.

LAUNAY, B., BURÉ, J., and PRADEN, J. 1977. Use of Chopin alveographe as a rheological tool. I. Dough deformation measurements. Cereal Chem. 54: 1042-1048.

MAES, E., and PIROTTE, A. 1957. Der Alveograph von Chopin: Konstante oder veränderliche Wasserzugabe. Getreide u. Mehl 7(1): 1-2.

MARCELLE, A. 1955. Kritische Betrachtungen zum Problemme der Mehlqualitätsklassifizierung durch den W-Wert des Alveographen. Getreide u. Mehl 5(9): 65-69.

NURET, H., SARAZIN, J., and BERRIER, M. 1970. Obtention d'alvéogrammes à P constant. Bull. de. l'Ecole Française de Meunerie, No. 235, 34-39.

SCHOFIELD, R.H., and SCOTT BLAIR, G.W. 1933. The relationship between elasticity and plastic strength of soft material as illustrated by some mechanical properties of flour dough. II. Proc. Roy. Soc. A. 139-557.

SCOTT BLAIR, G.W., and POTEL, P. 1937. A preliminary study of the physical significance of certain properties measured by the Chopin extensimeter for testing flour doughs. Cereal Chem.: 257-262.

TCHETVEROUKHINE, B. 1947. Les Alvéogrammes Chopin à matière sèche constante et à matière sèche variable. Bull. de l'Ecole Francaise de Meunerie, No. 102, 217-224.

TCHETVEROUKHINE, B. 1948. Les Alvéogrammes Chopin à matière sèche constante et à matière sèche variable. (Suite et fin) Bull. de l'Ecole Française de Meunerie, No. 103, 25-29.

YAMAZAKI, W.T. 1953. An alkaline water retention capacity test for evaluation of cookie baking potentialities of soft winter wheat flours. Cereal Chem. 30: 242-246.

WEIPERT, D. 1981. Teigrheologische Untersuchungsmethoden - ihre Einsatzmöglichkheiten in Mühlenlaboratorium. Getreide, Mehl. u. Brot 35(1): 5-9.

DO-CORDER AND ITS APPLICATION IN DOUGH RHEOLOGY

Seiichi Nagao
Research Center
Nisshin Flour Milling Co., Ltd.
Ohi-machi, Saitama 354, Japan

INTRODUCTION

Several rheological and chemical measurements currently employed on dough serve as indices which, when properly interpreted, enhance the probability of satisfactory endperformance. To date, however, none of these is capable of fully predicting the bread making performance of a dough with or without additives. The reason is assumed that all of them have been demonstrated at room temperature or 30°C.

Using a Brabender Do-Corder fitted with a closed mixing bowl and raised the temperature gradually to almost 100°C, we could reproduce the rheological properties of the bread dough in the early stage of baking which is the most important stage from the standpoint of the changes in flour components.

BRABENDER DO-CORDER

The Brabender Do-Corder is an instrument for testing the rheological behavior of cereal and other food products. This instrument is a torque rheometer which can be used with many interchangeable measuring heads. By selecting the appropriate measuring heads it is possible to test the quality of flour, the processing properties of dough at different mixing intensity, the influence of formulation and mixing, and the extrusion properties of alimentary pastes,

Whole view of the Brabender Do-Corder.

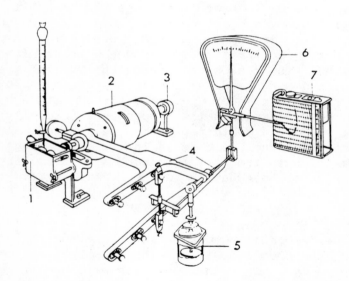

Diagram showing the principle of the Brabender Do-Corder. The measuring principle is similar to the one of Farinograph, however the drive is seven times as strong.   1, Measuring head;   2, free swinging dynamometer;   3, ball bearings;   4, lever system;   5, oil damper;   6, scale system;   7, recording device.

starches and other food products. The measuring
principle is similar to the one of the well-known
Farinograph, however the drive is seven times as
strong. Therefore, the instrument can be used as a
drive unit for all mixer and extruder measuring
heads. The speed of the instrument is stepless
adjustable from 5 to 250 rpm. In addition it is
supplied with a step switch for 10 fixed speeds.
When the measuring head is connected with a dynamometer, it is a precise instrument for the linear
measurement and recording of torque required to
shear-mix the sample in the measuring head at the
specific test temperature ranged from 40° to 300°C.
Through a unique lever arm system the dynamometer
transmits its reaction torque to both a scale and
a strip chart recorder.

A Brabender Do-Corder, with an almost completely
enclosed mixing bowl, was operated under the
following conditions: mixing speed, 60 rpm; setting
of the scale head, X1; position of buckle connector,
1:1; and chart paper speed, 1 cm/min. The measuring
ranges of the torque were shifted by adjusting a
lever from position 0 to position 5. The temperature
of the mixer was controlled at 35-110°C. The temperature of the test material was measured with a
thermocouple attached to the mixer and a continuous
recording chart. The correct amount of sample (70g
for flour and starch and 50g for vital gluten) was
put into the rotating mixer, and mixing was continued.
After 12min of mixing, the temperature was raised
gradually to almost 100°C (Tanaka et al 1980).

EFFECT OF OXIDANTS ON DO-CORDER CURVES

The changes in curve characteristics caused by
the various water absorption levels in doughs with
and without potassium bromate are illustrated in
Fig. 1. All curves tended to show either a double
rise or two peaks that closely coincided with the
temperatures of 75 and 85°C.

As the absorption level was decreased, the
major peak shifted from 75 to 85°C. With 1200ppm
bromate, however, the first peak at 75°C was clearly

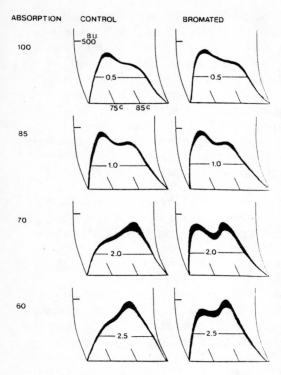

Fig.1. Do-Corder curves for control and bromated doughs at various absorption. The figure under each curve stands for the lever position of the instrument.

seen at any absorption. Accordingly, the effect of bromate was to produce characteristic curves with two clear peaks. Because this was especially true at 70% absorption, 70% was employed thereafter in this study. This level is also very close to that used in commercial baking.

The amount 1200ppm of bromate is greater than the level employed in the baking industry. However, the effects of bromate and of other oxidants could not be clearly detected by the Do-Corder unless oxidant levels were greater than 500ppm. The amount 1200ppm was arbitrarily chosen to accentuate the effects of the oxidants.

The Do-Corder curves on the left side of Fig.2 show the effects of the various oxidants on doughs at 70% absorption. Both iodated and ascorbated dough gave one clear peak at 85°C and 75°C, respectively and a small rise at 75°C and 85°C, respectively. In the presence of 1200ppm N-ethylmaleimide (NEMI), however, all curves were similar; they all showed a prominent peak at 85°C and a slight rise or a plateau at 75°C. These results imply that protein plays an

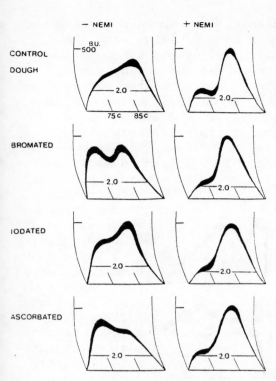

Fig.2. Effect of N-ethylmaleimide on Do-Corder curves for control and oxidized dough. The figure under each curve stands for the lever position of the instrument.

important role in forming the peak at 75°C. The peak at 85°C in the curves is attributed to the gelatinization and swelling of the starch. The consistency decreases as the temperature rises above 85°C and as mixing is continued.

INTERPRETATION OF DO-CORDER CURVES

To improve our understanding of the Do-Corder and consequently to increase its usefulness, we conducted experiments designed to identify the major components of flour responsible for the formation of the two peaks at 75 and 85°C. For this purpose, the effects on Do-Corder curves of starch damage and premixing in various mixers were investigated in the presence of reagents such as urea, salt, and N-ethylmaleimide(NEMI). Do-Corder curves for starch and vital gluten isolated from the flour and for defatted flour were also obtained (Endo et al 1981).

The protein fraction of flour appeared to be associated with a peak at 75°C, because urea (a protein-dispersing agent) and N-ethylmaleimide (a sulfhydryl-blocking agent) affected this peak.

Starch was the main flour component causing the increase in viscosity at 85°C, judging from the evidence that the damaging of starch affected it. Fig.3. shows Do-Corder curves for doughs with different levels of starch damage in the presence and absence of bromate. The curve for untreated flour showed a clear peak at 85°C in the absence of bromate. As the period of vibrating ball-mill treatment increased, the major peak (higher consistency) shifted from 85 to 75°C. Starch damage, however, had little effect on Do-Corder curves in the presence of bromate.

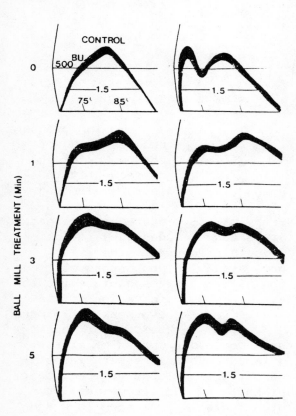

Fig.3. Do-Corder curves for ball-milled flours in the absence(left) and presence(right) of bromate. The figure under each curve stands for the lever position of the instrument.

Fig.4 shows Do-Corder curves for doughs from flour defatted either with petroleum ether or with water-saturated butanol. Extraction of free lipid from the flour with petroleum ether had little effect on the curves for normal or bromated doughs.

Extaction of total lipid with water-saturated butanol produced a drastic effect on the curves for all the samples, so that only a peak at 75°C was observed

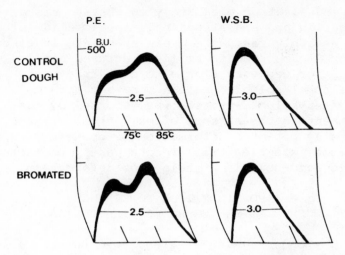

Fig. 4. Do-Corder curves for flours defatted with petroleum ether (P.E.) or water-saturated butanol (W.S.B.). The figure under each curve stands for the lever position of the instrument.

in each curve. This means that bound lipid plays an important role in determining the Do-Corder characteristics of a flour.

APPLICATION TO THE STUDIES OF THE STRUCTURAL CHANGES IN DOUGH PROTEIN

Do-Corder can be applied in many ways.

A study was designed as a visual examination of the structural changes produced in dough protein by mixing the Do-Corder in the presence and absence of the oxidants at elevated temperature (at 75 and 85°C), where distinct changes occured in Do-Corder curves (Nagao et al 1981).

Protein from flour and doughs mixed with a Do-Corder in the presence and absence of potassium bromate, potassium iodate and L-ascorbic acid was fractionated according to solubility into water-, salt-, alcohol-, acetic acid-soluble protein fractions and insoluble residue protein.

All fractions were freeze-dried and subjected to

scanning electron microscopy to observe visually the changes in protein structure. Acetic acid-soluble and insoluble residue protein are alike in structure, but the former was thermally denatured easily, while the latter was very stable to heat treatment. Salt- and alcohol-soluble protein were not deformed, but the water-soluble protein was deformed by heat treatment in the absence of oxidant. Oxidants generally promoted deformation of protein structure with the exception that bromate partly protected acetic acid-soluble protein from deformation.

A POSSIBLE TOOL TO EVALUATE THE BREAD-MAKING PROPERTIES OF A DOUGH

Do-Corder curves distinguished between a bromated dough (two peaks in curve at 75 and 85°C), a dough with added ascorbate (one peak at 75°C), and control dough with no additive (one peak at 85°C). These three doughs produced loaves with good, medium and poor volumes, respectively as shown in Fig. 5.

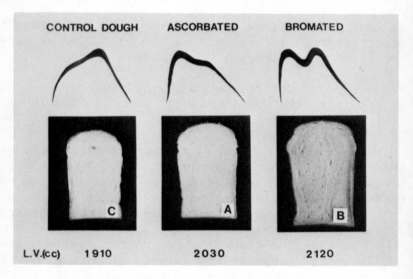

Fig. 5. Comparison of Do-Corder curves at 70% absorption and bread-making performances by the straight-dough method. C=control, A=ascorbated, B=bromated, L.V.=loaf volume.

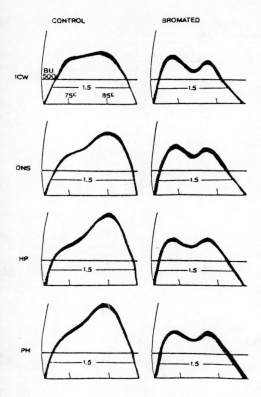

Fig.6. Effect of bromate on Do-Corder curves for flours with different baking quality. 1CW=No.1 Canada western red spring, DNS=dark northern spring, HP=hard red winter high protein, PH=Australian prime hard.

Do-Corder curves were somewhat different among flours with different breadmaking performance. In the presence of bromate, however, all curves (Fig.6) were alike and the quality of breads was much improved for all flours (Nagao et al 1981).

Forty-one compounds were tested with the Do-Corder in the presence of ascorbic acid. Nineteen gave two peaks in their Do-Corder curves and were thus similar to bromate in their curve (Table I).

Aspartic acid, glutamic acid, cystine, and potassium bitartrate were randomly chosen for straight-dough and sponge-dough baking tests. When incorporated in a dough with ascorbic acid, each of them improved to some extent the quality of bread baked by the straight-dough method (Table II).

Bread baked from a dough containing cystine and ascorbic acid by the sponge-dough method showed the same baking performance as bread baked from bromated dough.

Therefore, it can be said that the Do-Corder is a promising tool for predicting the baking

TABLE I
Effects of Compounds on the Do-Corder Curve

Compound	Do-Corder Curve[a] 75°C Shoulder	Two Peaks in Presence of Ascorbic Acid
Amino acid		
Glycine	+[b]	−
Alanine	+	+
Glutamic acid	+	+
Aspartic acid	+	+
Methionine	+	+
Histidine	+	−
Cystine	+	+
Cysteine	−	−
Tryptophan	−	−
Leucine	−	−
Arginine	−	−
Organic acid salt		
Malic acid	+	−
Ferrous lactate	+	+
Sodium pantothenate	+	−
Potassium bitartrate	+	+
Ferric citrate	+	−
α-Ketoglutaric acid	+	+
Succinic acid	+	−
Ferric ammonium citrate	−	−
Monosodium succinate	−	−
Monosodium citrate	−	−
Citric acid	−	−
Fumaric acid	−	−
Inorganic salt		
Sodium metaphosphate	+	+
Ammonium alum	+	+
Potassium alum	+	+
Magnesium chloride	+	−
Sodium sulfate	+	−
Sodium tripolyphosphate	−	−
Others		
Caffeine	+	+
EDTA[c]	+	+
Tannic acid	+	−
Albumin	+	+
D-Sorbitol	+	−
Nicotinic acid	+	+
Nicotinamide	+	−
Linoleic acid	+	+
Stearic acid	+	+
GDL[d]	+	+
Pectin	+	−
Xylose	−	−

[a] All compounds gave a peak at 85°C.
[b] + = Present, − = absent.
[c] Ethylenediamine tetraacetic acid.
[d] Glucono delta lactone.

TABLE II

Effects of Various Compounds on Bread-Making Properties Using the Straight-Dough Method

	Dough or Bread										
	A	B	C	D	E	F	G	H	I	J	K
Compound ppm											
Potassium bromate	...	10	...	...	...	...	...	...	...	...	...
Ascorbic acid	...	...	6	6	6	6	6	6	6	6	6
Aspartic acid	...	...	...	15	30	...	...	...	...	...	...
Glutamic acid	...	...	...	...	...	15	30	...	...	...	...
Cystine	...	...	...	...	...	...	...	15	30	...	...
Potassium d-bitartrate	...	...	...	...	...	...	...	...	...	15	30
Bread property											
Loaf volume (cc)	1,890	2,030	1,950	2,010	2,000	2,010	2,020	2,010	2,020	1,970	1,970
Loaf volume rating[a] (1–15)	11.1	13.3	12.5	13.1	13.0	13.1	13.2	13.1	13.2	12.7	12.7
Crust color rating (1–10)	7.5	7.7	7.7	7.7	7.6	7.4	7.5	7.8	7.8	7.5	7.5
Crust characteristic rating (1–15)	9.8	11.6	11.0	11.6	11.0	11.3	11.3	11.6	11.9	11.4	11.3
Crumb color rating (1–10)	6.4	7.5	7.5	7.4	7.4	7.7	7.7	7.5	7.4	7.5	7.5
Grain rating (1–20)	12.4	15.4	15.0	15.2	15.2	15.0	15.0	15.0	15.4	15.2	15.2
Texture rating (1–20)	13.2	15.4	15.2	15.2	15.2	15.2	15.4	15.0	15.0	15.0	15.0
Flavor rating (1–10)	7.4	7.5	7.5	7.4	7.4	7.5	7.5	7.5	7.5	7.4	7.4
Total[b]	68.6	78.4	76.4	77.6	76.8	77.2	77.6	77.5	78.2	76.7	76.6

[a] In all rating systems, 1 = least desirable.
[b] 100 possible.

performance of a dough containing certain improver additives.

EFFECT OF FERMENTATION ON DO-CORDER CURVES

The Do-Corder curve of a fermenting leavened dough showed two peaks (at 75 and 85°C) and was similar to the curve for a bromated dough(Fig.7).

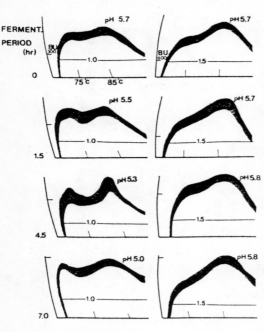

Fig.7 Do-Corder curves for leavened (left) and unleavened (right) dough after various periods of fermentation. The number under each curve stands for the lever position of the instrument.

The addition of prefermented dough to a freshly prepared one produced a Do-Corder curve with two peaks and improved the baking performance as evaluated by the sponge-dough baking procedure.

Substances produced by yeast during fermentation appeared to play a major role in modifying the physical properties of a dough. The change in Do-Corder curve and the improvement in bread-making potential associated with the fermentation step were not due solely to the change in dough pH; the addition of acetic acid to an unleavened dough to lower its pH to the level of a leavened one showed little effect on its properties (Nagao et al 1981).

FACTORS AFFECTING THE RHEOLOGICAL PROPERTIES OF DOUGHS IN TERMS OF DOUGH DEVELOPMENT

A Do-Corder was used to study factors affecting the rheological properties of doughs in terms of dough development. The major peak of the Do-Corder curve for a dough, developed in a mixograph with a 35-g bowl before heat treament, shifted from 85°C to 75°C, compared with that for a control flour. The shift was influenced by mixing time and absorption level. The shift in this peak was promoted by an increase in the water absorption level(Fig. 8).

There was a rapid decrease in the content of sulfhydryl and free lipid extracted with ethyl ether at an early stage of mixing.

However, the decrease in sulfhydryl content for a control flour after heat treatment in the Do-Corder was small compared with that for a premixed dough.

Free lipid in a premixed dough increased greatly

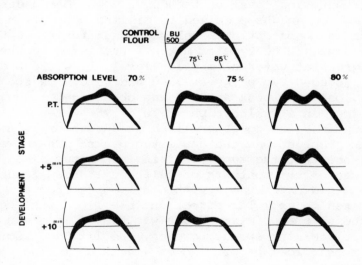

Fig. 8. Do-Corder curves for a control flour and powdered freeze-dried doughs premixed in a mixograph for the appropriate times at various absorption levels(70, 75, and 80%). Doughs were mixed until they reached maximum consistency, and then for another 0(peak time), 5, and 10 min.

as a result of heat treatment in the Do-Corder, despite a slight decrease in free lipid observed in a control flour after heating. Flour protein was responsible for the rheological changes of doughs during prolonged mixing, judging from the changes in shlfhydryl content and lipid binding.

On the other hand, little response was shown in Do-Corder curves for doughs containing N-ethylmaleimide(Endo at el 1984).

EFFECT OF MIXING APPARATUS AND MIXING SPEED ON THE RHEOLOGICAL AND COMPOSITIONAL PROPERTIES OF HEATED DOUGH

Factors affecting the formation of two major peaks at 75 and 85°C in the Do-Corder curve for a dough developed in various mixers prior to heat treatment were studied. The shape of these peaks depended on both the type of mixing apparatus and the absorption level of the dough. The Do-Corder curves for doughs premixed to miximum consistency at a constant absorption level showed different patterns depending on the type of mixer. Premixing in the Mixograph at water absorption level of 70, 75 and 80% produced two peaks at 75°C and 85°C. On the other hand, only one peak at 85°C was observed for doughs premixed in a Farinograph at 70 or 75% water absorption; the dough premixed at 80% absorption on a Farinograph showed the usual two peaks. There were evidence that the type of mixing apparatus also influenced the contents of sulfhydryl groups and of free lipid extractable with ethyl ether from a premixed dough. The effect on free lipid was accentuated when a premixed dough was heated in the Do-Corder. Mixing speed, however, had little effect on the rheological properties of doughs as tested by the Do-Corder.

SUMMARY

The Brabender Do-Corder fitted with a closed mixing bowl and operated at a temperature as high as that in the early stage of baking is an useful instrument for studying the improving mechanism of oxidants in bread dough and a promising tool for predicting the baking performance of a dough containing certain improver additives. The protein and starch fractions of flour appeared to be associated with two peaks at 75 and 85°C in the Do-Corder curve, respectively. It is applicable to test the rehological properties of a fermenting leavened dough and to study factors affecting the rheological properties of doughs in terms of dough development, too. Interactions of water absorpeion, sulfhydryl level and free lipid content in heated dough as well as effects of mixing procedures on the rheological properties of heated dough were studied.

ACKNOWLEDGMENTS

I gratefully acknowledge the advice and assistance of my co-workers, K. Tanaka and S. Endo in the preparation of the manuscript.

LITERATURE LITED

Endo, S., Nagao, S., and Tanaka, K. 1981. Interpretation of Do-Corder curves. Indentification of flour components influencing curve characteristics. Cereal chem. 58: 538

Endo, S., Tanaka, K., and Nagao, S. 1984. Do-Corder studies on dough development. I. Interactions of water absorption, sulfhydryl level, and free lipid content in heated dough. Cereal chem. 61: 112

Nagao, S., Endo, S., and Tanaka, K. 1981. Scanning electron microscopy studies of wheat protein fractions from doughs mixed with oxidants at high temperature. J. Sci. Food Agric. 32:235

Nagao, S., Endo, S., and Tanaka, K. 1981. Effect of fermentation on the Do-Corder and bread-making properties of a dough.
Cereal chem. 58:388

Nagao, S., Endo, S., and Tanaka, K. 1981. The Do-Corder as a possible tool to evaluate the breadmaking properties of a dough. Cereal chem. 58: 384

Tanaka, K., Endo, S., and Nagao, S. 1980. Effect of potassium bromate, potassium iodate, and L-ascorbic acid on the consistency of heated dough. Cereal chem. 57:169

DYNAMIC RHEOLOGICAL TESTING OF WHEAT FLOUR DOUGHS

J. M. Faubion[1], P. C. Dreese[1] and K. C. Diehl[2]

[1]Department of Grain Science and Industry, Kansas State University, Manhattan, Kansas 66506.

[2]Department of Agricultural Engineering, Virginia Polytechnic Institute and State University, Blacksburg, Virginia 24061.

Of all the chemical and physical aspects of wheat flour dough, accurate, objective assessment of "texture" is one of the most important (Faubion and Diehl, 1984). The mechanical energy, or work of mixing transforms flour and water into a viscoelastic dough. During subsequent processing, that dough is subjected to more work: kneading, sheeting, dividing, rounding, molding, all of which modify its texture (Heaps et al 1967). Simply doing nothing, allowing the dough to rest or relax will change its textural properties. An important corollary to this fact is the fact that the rheological properties of a dough can have profound effect on the quality of the final product. Thus, too much or too little work (or too much or too little relaxation) at many stages of bread production can reduce the quality of the baked product. Obviously, then, the accurate assessment of the textural characteristics of dough is central to the quality of the final product. In another and more basic sense, knowledge of the fundamental rheological properties of wheat flour doughs is critical to our understanding of the chemistry of the dough system.

This fact has been appreciated by cereal chemists for a number of years and has spurred the development of a number of widely used texture analysis instruments specifically designed for dough (Mixograph, Farinograph, Extensograph, Alveograph). The great amount of research done with these instruments testifies to their worth in providing information about dough's behavior in specific circumstances and geometries. Most, if not all of the above instruments fall short of measuring the fundamental rheological properties of the materials they test. This is because the results they give are dependent on sample size and/or shape. In addition, they often impose deformations on the sample large enough to cause changes in the mechanical properties of the dough due to the test itself.

A related group of techniques, the dynamic force-deformation tests have been used for a number of years by rheologists interested in the mechanical properties of polymers such as plastics. Since these tests have the potential to evaluate a number of rheological properties of foods including dough (Rao, 1984), it isn't surprising that they have drawn the attention of an increasing number of food rheologists. Dynamic sinusoidal stress-strain testing is becoming one of the more common rheological testing methods used to determine viscoelastic properties of food materials. Dynamic measurements are particularly useful in measuring short time or high rate rheological behavior as well as behavior at very low deformations and strains.

THEORETICAL CONSIDERATIONS

As with all such tests, the basis of the dynamic stress-strain technique is the relationship between stress, strain, and time. In the tests, samples are acted upon by one or more forces, body forces (gravity, inertia) or surface forces, acting at the surface of the test piece. The deformation resulting from the action of such a force is a function of both the force itself and the area over which that force is distributed. This is defined as stress and calculated as

$$\tau = F/A \qquad (1)$$

where τ = stress, F = force acting upon the sample and A, the area over which the force acts. Stress, therefore, is measured or calculated in Pascals (Newtons/N^2).

A stressed material will often deform and the definition is, in rheological terms, strain. Simple shear strain is illustrated in Fig. 1 and, for this geometry, is calculated as

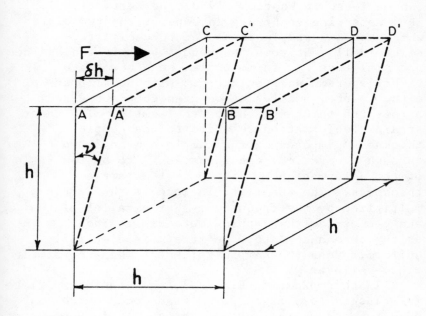

Figure 1. Cubical test sample in simple shear. Force, F acts in the direction indicated along plane A,B,C,D, causing a deformation, h. The resulting position of the plane is now A',B',C', D'. Reproduced from Faubion and Diehl (1984), with permission.

$$\gamma = \delta l/h \qquad (2)$$

l is the distance moved by one plane surface relative to the other.

The brief review of the equations and measured quantities involved in dynamic testing which follows uses the parallel plate geometry shown in Fig. 2a as an example. This technique is related to the assortment of dynamic measurement techniques which are found in the literature on dynamic viscoelastic testing of wheat doughs. It is not intended to represent the wide variety of geometries which exist. More detailed discussions of these geometries and the theory involved in dynamic testing can be found in Whorlow (1980) and Ferry (1980).

Referring to Figure 2a, one of the two plates which bound a sample is caused to oscillate, sinusoidally, at some frequency, (w), in radians/sec and amplitude, d, in mm while the other plate (ideally) remains stationary. With no slippage at either plate, a deformation gradient is created across the thickness (h) of the sample. The shear strain (deformation relative to the sample thickness) imposed on the sample can be shown, through theoretical considerations, to be the ratio of the deformation amplitude to the sample thickness. The shear strain rate is the strain multiplied by the frequency. For parallel plates and geometries that approximate simple shear, such as the cone-and-plate, the strain is esentially uniform across the sample thickness when the sample behaves linearly.

In this example (Fig. 2a), the stationary plate is attached to a rigid force transducer which is used to measure the force response, f, in Newtons of the sample. The shear stress which results from the force being distributed over the sample area (l x w) is also uniform over the sample thickness. If the sample is linearly viscoelastic, the force response will vary sinusoidally as shown in Fig. 2b, at the same frequency as the applied deformation. It may lag the deformation by some amount, ϕ, the phase angle, (in radians). The sinusoidally varying shear stress and shear strain are represented mathematically as:

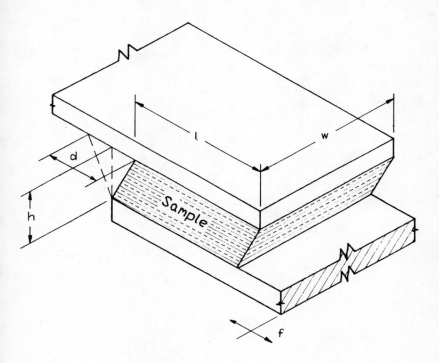

Figure 2a. Parallel plate geometry for dynamic testing.

$$\tau = (f/l\ w)\sin(wt-\phi) = \tau_o \sin(wt-\phi) \qquad (3)$$

$$\gamma = (d/h)\sin(wt) = \gamma_o \sin(wt) \qquad (4)$$

where τ_o and γ_o are shear stress and shear strain amplitudes respectively. For convenience, the stress and strain can be expressed in complex variable notation. The ratio of the complex stress and complex strain gives the complex modulus, (G*) which can be expressed as

$$G^* = (\tau_o/\gamma_o)\ (\cos\phi - i\sin\phi) \qquad (5)$$

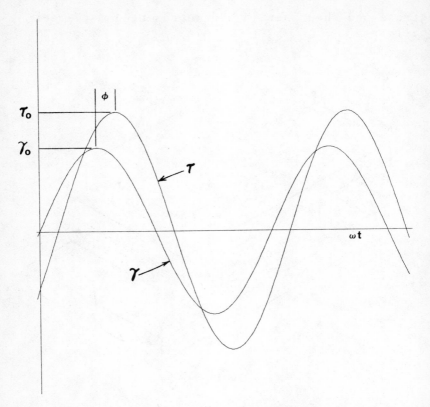

Figure 2b. Sinusoidal signal output from the force and deformation transducers. Indicated values are defined in the text.

The storage modulus (G') and loss modulus (G") are the coefficients of the real and imaginary components respectively of the complex modulus. Therefore,

$$G' = (\tau_o/\gamma_o)\cos\phi \qquad (6)$$

$$G'' = (\tau_o/\gamma_o)\sin\phi \qquad (7)$$

The absolute value of the complex modulus is given by

$$|G^*| = (G'^2 + G''^2)^{1/2} \qquad (8)$$

As Ferry (1980) points out, reference to G* is an abbreviated way of referring to G' and G" simultaneously. In practice, f, d, and ϕ can be measured at a variety of frequencies and strains. G' and G" are usually reported or modelled as a function of frequency. These values can also be used to determine other linear viscoelastic properties of materials such as wheat flour doughs. Commercially available dynamic testing instruments commonly yield G', G" and related parameters directly. The phase angle provides a simple index of the relative elastic or viscous nature of a material. This property is sometimes expressed as the loss tangent (tan ϕ) which is the ratio of the loss modulus to the storage modulus (G"/G'). This result can be obtained by dividing equation 6 by equation 5. Note that as the phase angle approaches 0, a material is behaving more like an elastic solid. Conversely, as the phase angle approaches 1.57 radians or 90 degrees, a material is behaving like a Newtonian liquid.

TECHNICAL CONSIDERATIONS

It must be remembered that the equations discussed above apply to any geometry. For geometries other than parallel plate, the stress and strain would be calculated differently. Not accounted for in these equations are the effects of sample inertia. Under ideal conditions, the wave length of the shear wave transmitted through the sample is much larger than the sample thickness (h). In such a case, the stress and strain states would exist as predicted by theory. However, if the gap or sample thickness (h) is too large, the force response may be attenuated due to sample inertia. Ferry (1980) suggests that the sample density should be small compared to $G'/h^2(w/2)^2$ or G" divided by the same expression for inertial effects to be negligible. The moduli of wheat flour doughs are usually of sufficient magnitude and the testing frequencies low enough that sample inertia effects are negligible.

Another consideration which is of particular importance in dough rheology is the condition of the surface of the dough exposed to the surrounding environment (ie, sample edges in the parallel plate geometry). Moisture loss and crust formation are common problems in many aspects of dough research. When this occurs at the exposed surfaces of dough samples being tested as illustrated in Figure 2a, significant stresses can be created. These produce forces not accounted for in the equations which are used to calculate the moduli. This will, then, result in incorrect results.

The two most common methods of minimizing dehydration are coating the exposed edges and/or humidifying the environment close to the exposed surfaces. Szczesniak et al (1983) cited attempts at minimizing dehydration and presented a vivid example of the effect that exposed surfaces can have. When comparison was made between a sample of dough coated with a low-viscosity silicone oil and a sample exposed to air, the storage modulus of the uncoated dough was roughly 10 times higher than that of the coated sample. As the authors were careful to point out, ideally, the humidity of the air surrounding the dough would be equal to the water activity of the dough. This, unfortunately, is often impractical, particularly if the effect of temperature on dough behavior is being explored.

In addition to moisture loss, the history of the dough immediately prior to testing must be considered. Because doughs are subject to relaxation and dehydration if not somehow preserved, they are usually tested soon after preparation. After mixing, the dough must be shaped into a form that is suitable for insertion into the plates of the testing apparatus. Often, after insertion, the dough must be further compressed in the test plates to obtain a prescribed sample thickness. The procedures are often unavoidable and can induce residual stresses in the sample that can influence subsequent measurements. To reduce the magnitude of this problem, samples are often allowed to "rest" before testing so that the residual stresses will relax. Various rest times from 5 minutes (Szczesniak et al, 1983) to 3 hours (Hibberd and Wallace, 1966) have been used. Navickis et al (1982) assessed the effects of rest times of 1, 3, and 5 hours. They found no further changes after 1 hour. In some instances, (P. C. Dreese, personal communication), it has been possible to sheet dough to be tested to thicknesses that closely approximate the desired thickness. This has allowed reductions in the amount of stress imposed on the sample during loading, and, therefore, the amount of relaxation time required.

The problem of changes in dough rheology due to relaxation may be of more importance than generally realized. If researchers wish to study the properties of dough at a particular time in the bread production scheme or, even at a particular time during or after mixing, it may well be faulty to allow long rest times before testing. Changes that would normally occur in dough during long rest or relaxation times may obscure any changes in rheology due to treatment etc. Put more directly, it may be difficult to sort out changes in textural properties due to relaxation from those due to experimental manipulation.

INSTRUMENTATION

Commercially available instruments for dynamic testing grew out of the demands of the polymer industry rather than the food industry. The most widely used of the instruments currently available, the Mechanical Spectrometer (Rheometrics Inc., Union, N.J.) has found use in research on dough rheology (Szczesniak et al, 1983). An alternate approach to istrumentation has been to assemble an instrument from commercially available transducers and associated electronics. This approach has been most common in the testing of foods, including doughs (Hamann, 1969, Hibberd and Parker, 1975, Hibberd and Wallace, 1966). In the past, this was due to the non-existance of commercially available instrumentation. Currently, it reflects the wide variety of foods (in terms of their form and ability to be shaped) that are being tested.

Since it involves small deformations and relatively high frequencies, the instrumentation for dynamic testing is somewhat complex. It can be considered to consist of five sections or systems: 1) a function generator, power amplifier and vibration exciter to provide the sinusoidal deformation, 2) a linear variable differential transformer to measure deformation, 3) a force transducer, 4) a test stand and sample holder of defined geometry and 5) a dual trace oscilliscope or computor to collect and process data. A schematic diagram and photograph of one such system, assembled in the author's (JMF) laboratory are shown in Figures 3 and 4.

RESEARCH FINDINGS

Since dynamic stress-strain analysis of dough is, relatively speaking, a new technique, the amount of research using it is not large. Thus, no concensus exists in the literature as to the dynamic rheological properties of dough. Likewise, no model exists of wheat flour dough based on the results of such tests. Still, a number of questions have been posed and then examined by dynamic, stress-strain analysis. Those to be considered below include: 1) non-linear behavior of dough, 2) the effects of varying water contents, 3) the effects of varying mixing time, 4) the effects of variations in strain amplitude and frequency and 5) differences in the dynamic rheological behavior of flour dough vs. gluten.

NON-LINEAR BEHAVIOR OF WHEAT DOUGHS

All of the technical as well as the theoretical discussion above assumed that the material being tested behaves linearly. The theory of linear viscoelasticity assumes that the properties of a material, such as G' and G'', are a function of time or rate only. Hence, they are not affected by the

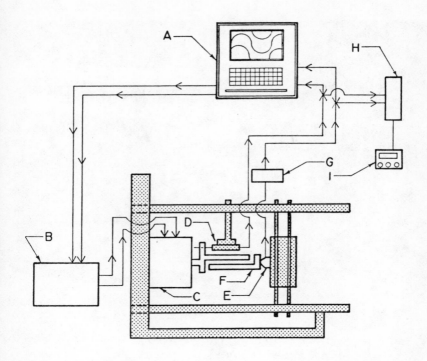

Figure 3. Schematic diagram of an instrument for dynamic sinusoidal stress-strain testing. A) microcomputer used as a signal generator and digital oscilliscope, B) power amplifier, C) vibration exciter, D) deformation transducer, E) force transducer, F) rigid test stand, G) transducer coupler, H) amplifier, I) phase to voltage converter. Reproduced from Faubion and Diehl (1984) with permission.

magnitude of the stress or strain imposed on them. Most viscoelastic materials behave linearly only at low strain levels but the level at which non-linearity occurs can be quite variable. Non-linear behavior can also be highly dependent on material structure and composition. Linearity can be determined by varying the strain level during testing and observing the constancy of the calculated moduli. Also, the force response of the material to the

Figure 4. Photograph of the dynamic testing apparatus diagrammed above in operation.

sinusoidal strain itself should remain sinusoidal.

When a material ceases to behave linearly, then more complex theory and models may be necessary to describe its behavior beyond the linear range (Bird et al. 1977). Unless non-linearity is being specifically modeled, however, the moduli are often calculated using linear viscoelastic theory on the basis of simplicity. This has usually been the case in the dynamic testing of wheat flour doughs.

The onset of non-linearity is of particular interest since it may identify some change in the molecular behavior of the material due to deformation. For example, Baird (1981) hypothesized that hydrogen bonds and hydrophobic interactions were responsible for the linearity (up to 30% strain) of doughs containing 20 and 25% soy isolate. Doughs of 15% isolate concentration did not behave linearly and the closer proximity of the molecules at higher protein concentration was cited by the authors as a contributing factor in this behavior.

The linearity of wheat dough behavior has been examined by several researchers (Hibberd and Wallace, 1966, Smith et al. 1970, Hibberd and Parker, 1975, Matsumota, 1979, Szczesniak et al. 1981, Navickis et al. 1982). There appears to be little agreement in the literature as to the strain level at which linear behavior ceases. This is not surprising in view of the wide variety of wheat cultivars used to prepare the doughs, the variety of mixing methods used and the differences in testing procedures. It is clear, however, that wheat doughs cease to behave as linear viscoelastic materials at very low strains.

Hibberd and Wallace (1966) found that dough made from the Austrialian wheat variety Mendos behaved linearly up to a strain of 0.0022. Nonlinearity (dependence on both stress and strain magnitude) was observed from this strain to a strain of 0.04. No statement as to the moisture content of the dough was found but reference was made to the dough being mixed to optimum development in a Farinograph.

EFFECTS OF FREQUENCY AND STRAIN AMPLITUDE

Frequency dependence has been the subject of a number of studies. G' and G" of flour doughs are both frequency dependent and increase with increasing frequency (Hibberd and Wallace, 1966, Smith et al. 1970, Cumming and Tung, 1977 and Szczesniak et al. 1983). For gluten-starch doughs, frequency dependence increases as protein/starch ratio decreases (Hibberd, 1970b) (Fig. 5).

The general result of nonlinear behavior is a decrease in the magnitude of the moduli as strain amplitude increases (Smith et al. 1970, Szczesniak et al. 1983). Strain amplitude also affects the relative magnitude of the moduli. At low strains (less than 0.1) G' exceeds G" while at strains greater that 0.1, G" is greater than G'. This behavior has been noted in both dynamic (Szczesniak

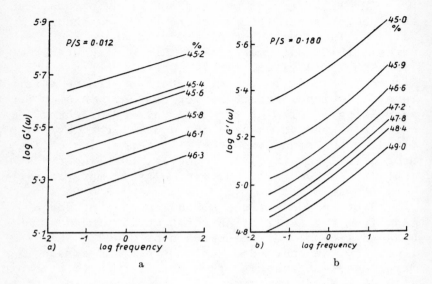

Figure 5. Influence of water content on the storage modulus of protein-starch-water systems at varying protein/starch ratios. Reproduced from Hibberd (1970b) with permission.

et al. 1983) and tensile tests, (Funt Bar-David and Lerchenthal, 1975). At low strains, flour doughs behave more as solids while at higher strains, this changes to more fluid like behavior. Funt Bar-David and Lerchanthal (1975) characterized this change as being from a viscoelastic solid to an elastoviscous liquid. Cumming's data, which demonstrated that the viscous response of a dough increases as energy input increases supports these characterizations.

EFFECTS OF VARYING MIXING TIME

Bohlin and Carlson (1982) investigated the dynamic rheological properties of bread flours that differed in strength and concluded that both moduli (G' and G") were dependent on mixing time (Fig. 6).

In addition, this dependence was shown to differ among the various flours tested. It should be noted that the test conditions (frequencies of 0.08-20.9 rad/sec at a strain of 0.37) likely resulted in non-linear behavior. It is interesting to speculate as to whether or not the observed differences were due to differences in the flours' optimum mix time or stability to over mixing.

The fact that optimum mixing times exist and may well vary among different flours is often slighted or, at worst, over-looked, in studies of the dynamic rheological properties of doughs. Bohlin and Carlson's results suggest that such an oversight is a significant mistake. Reliance on mixing procedures that establish a fixed mixing time or mixed consistancy (eg. in BU) for differing flours or treatments may well make any comparisons of the dynamic properties of the resulting doughs invalid. A similar argument can be made against the use of fixed water absorption levels.

EFFECTS OF WATER CONTENT

Obviously, after the flour itself, water is the most critical factor in the creation of a viscoelastic dough. A number of studies have examined the effects of varying water content on the dynamic viscoelastic behavior of wheat flour doughs (Hibberd, 1970a, Hibberd and Parker, 1975, Navickis et al. 1982, Smith et al. 1970). The consistant conclusion of this work was that both the storage and loss moduli of wheat flour dough decrease with increasing water content. The same effect was observed for doughs composed of wheat starch plus gluten (Hibberd 1970b) (Figs. 4 and 7). Hibberd and Parker's (1975) research showed that, within the ranges of their experimental observations (0.32-32 Hz and 44-49% water) there was no interaction between the effects of frequency (see above) and dough water content. The effects were, therefore, considered to be separable. In contrast, Navickis et al. (1982) demonstrated an interaction between the affects of testing frequency and dough water

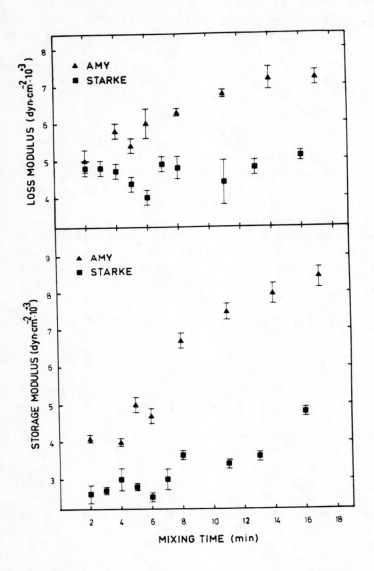

Figure 6. Dynamic loss and storage moduli as a function of mixing time for two flour varieties (Amy and Starke). Error bars present the standard deviation from means of at least three measurements. Reproduced from, Bohlin and Carlson (1981), with permission.

content (Figs. 8 and 9).

EFFECTS OF PROTEIN CONTENT

Most, if not all of the research investigating the effects of differences in protein content have been carried out using combinations of wheat starch and wheat gluten combined to achieve a specific protein content in the final blend. Since these are not doughs in the same sense that combinations of native flour plus water are, it is more accurate to refer to this research as the effects of protein/starch ratio.

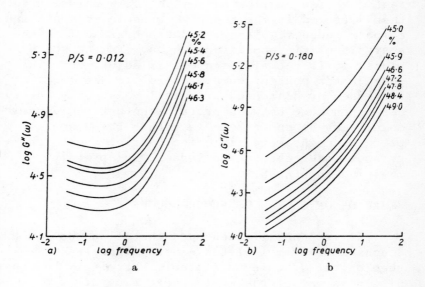

Figure 7. Influence of water content on the loss modulus of protein-starch systems at varying protein/starch ratios. Reproduced from Hibberd (1970b) with permission.

Smith et al (1970) observed that, at constant water levels, G' increased with protein content.

In fact both moduli increased with protein content and the effect was most pronounced at higher protein levels. Hibberd (1970b) found basically the same relationship and demonstrated that the frequency dependence of both moduli increased as the protein/starch ratio in the "dough" increased (Fig. 10). Finally, Navickis et al. (1982), who tested doughs over a wide frequency range (1-70 rad/sec) also demonstrated that decreasing protein content results in lower values for both storage and loss moduli.

It is important to use caution in the interpretation of these data for at least two reasons. First, native flour doughs clearly behave differently from pure gluten or blends of isolated gluten and starch. An interesting study would be to test a series of flours from the same wheat variety that differed in their native protein contents and compare those results with data obtained from the analysis of gluten-starch doughs. In the second place, it is not always clear if adjustments in dough water content were made to account for increased or decreased protein levels. The observed correspondence between protein content and dynamic moduli may be due to the fact that the higher protein doughs were, in effect, drier due to the greater water absorption capacity of protein relative to starch.

BEHAVIOR OF GLUTEN VS FLOUR DOUGHS

It is generally conceded that flour doughs and doughs made from isolated gluten alone are, rheologically, quite different. Precisely how they differ is one of the most interesting areas of research using the dynamic stress strain technique.

The most startling difference uncovered by dynamic testing involves linearity in the behavior of gluten vs. flour doughs. Doughs composed of isolated gluten and water have been found to behave in a linear fashion to a much greater degree than have doughs produced from native, unfractionated flour. Studying the dynamic mechanical behavior of gluten plus starch blends, Smith et al. (1970) found

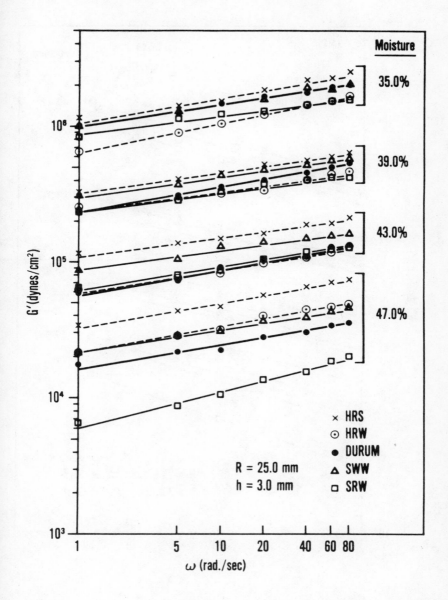

Figure 8. Storage modulus versus frequency for five wheat types over four moisture contents. Reproduced from Navickis et al. (1982) with permission.

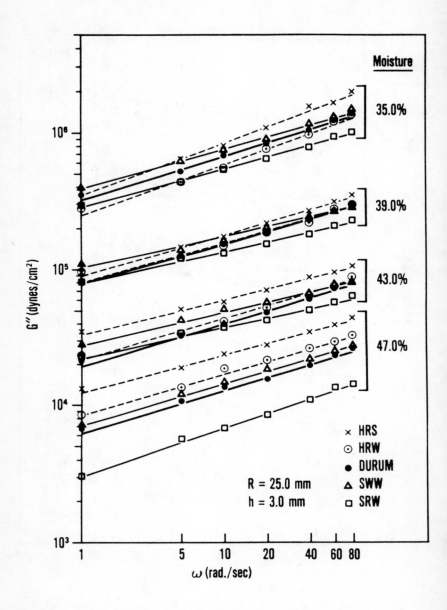

Figure. 9. As in Figure 8 but for loss modulus. Reproduced from Navickis, et al. (1982) with permission.

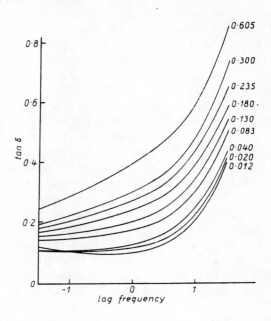

Figure 10. Loss tangent versus frequency for protein/starch systems. Protein/starch ratios are given at the extreme right of each curve. Reproduced from Hibberd (1970) with permission.

that the only doughs to give linear viscoelastic behavior at strains up to 0.021 were those composed of 100% gluten. Szczesniak et al. (1983) showed that increasing the gluten to starch ratio (45:55 to 100:0) in samples tested with the Rheometrics Dynamic Mechanical Spectrometer, increased the linear portion of plots of w vs. G'. Hibberd's dynamic analysis (1970b) of gluten plus starch doughs indicated that the starch components was not serving simply as an inert filler in the dough. It was, in fact, playing an active role in determining the dough's textural properties. Matsumoto (1979), using stress relaxation measurements on gluten and gluten plus added starch, concluded that starch granules were introducing non-linearity in gluten behavior.

The results above appear to suggest that hydrated gluten behaves quite linearly and that this behavior is very different from that exhibited by doughs created from native flour. Starch added to pure gluten appears to disrupt this linear behavior as well as to increase the dynamic stiffness of the resulting dough. Because of this behavior, an analogy between wheat dough and filled elastomers such as carbon black-filled rubber has been advanced as a model for dough structure. While this may be helpful in understanding the rheological behavior of dough, caution should be exercised in extending such analogies to molecular behavior. Doughs are much more complex in their molecular interactions than synthetic polymers. There is, as yet, no explanation of the precise reason for the observed differences between gluten and flour doughs. Pursuit of that explanation will, doubtless, be one of the more fruitful areas of research using the dynamic, sinusoidal stress-strain technique.

PRODUCT STUDIES

Due to the inherent complexity of the dough system, most of the research to date on dough rheology by the dynamic technique has been basic in nature. The focus has been on unyeasted flour-water doughs. Its intent has been to characterize that system and the effects of varying integral dough components such as water (Hibberd, 1970a) or protein (Hibberd, 1970b, Navickis et al. 1982).

Of late, the dynamic testing approach has been used to examine the rheological behavior of more complex doughs or processing conditions. LeGrys et al. (1981) showed that heating isolated gluten from $50°$ to $100°C$ resulted in roughly linear increases in both storage (G') and loss (G'') moduli. The authors concluded that during heating, changes in the elastic properties predominated and that this was consistent with an increase in the number of rheologically effective cross-links between peptides.

In a study extending that work, Schofield et

al. (1985) investigated the relationship between breadmaking quality and the rheological properties of laboratory-isolated, heated glutens. They demonstrated a simple, inverse relationship between loaf volume and storage modulus. Thus, as G' increased (due to heating), loaf volumes were lower. The same inverse relationship existed when commercially isolated gluten was tested. In this case, the slope of the regression line (loaf volume vs. G') was less steep. In both instances, the relative viscous and elastic properties of the protein were seen to change so that elastic behavior became excessive and damaged baking functionality.

A series of experiments on dough by Szczesniak (1983), to study frequency dependence (see above) of G' and G'', reported a frequency-specific discontinuity or "dip" in both values. The dip was more pronounced in G'. It could be eliminated by; coating exposed sample edges with oil, adding shortening plus emulsifier to the dough or increasing strain levels to 0.1 or greater. Addition of urea to the dough reduced the severity of the dip.

The authors' explanation for the discontinuity is related to what they term the "hydrodynamics of the starch component". It was postulated that the dip was the result of the induced breakdown of starch aggregates in the dough.

Cumming (1975) added 1-5% fat to gluten and observed consequent decreases in both G' and G''. Szczesniak et al. (1983) confirmed this finding. In extending the work to wheat flour doughs, they found that the presence of shortening further altered the dough's dynamic viscoelastic properties. A flour-water dough was found to be a visco-elastic solid (G' greater than G'') while a dough with shortening was a elasto-viscous liquid (G'' greater than G'). The speculation (Szczesniak et al. 1983), that this phenomenon may be dependent on the mode of fat incorporation is, as yet, unsupported.

LITERATURE CITED

Baird, D. G. 1981. Dynamic viscoelastic properties of soy isolate doughs. J. Texture Studies. 12:1.

Bird, R. C., Armstrong, R. C. and Hassager, O. 1977. Dynamics of polymeric liquids. Vol. 1. Wiley, New York, NY.

Bloksma, A. H. 1971. p. 524. In: Wheat Chemistry and Technology. Ed. Y. Pomeranz. American Association of Cereal Chemists, St. Paul, MN.

Bohlin, L. and Carlson T. L. G. 1980. Dynamic viscoelastic properties of wheat flour dough: Dependence on mixing time. Cereal Chem. 57:175.

Chattopadhyay, P. K., Hamann, D. D. and Hammerle, J. R. 1978. Dynamic stiffness of rice grain. Trans. ASAE. 21:786.

Clark, R. C. and Rao, V. N. M. 1978. Dynamic testing of fresh peach texture. Trans. ASAE. 21:777.

Cumming, D. B. 1975. Rheological and Ultrastructural Properties of Wheat Gluten. PhD Thesis, University of British Columbia.

Cumming, D. B. and Tung, M. A. 1977. Modification of the ultrastructure and rheology of rehydrated commercial wheat gluten. Can. Inst. Food Sci. Technol. J. 10:109.

Faubion, J. M. and Diehl, K. C. 1984. Dynamic testing of wheat flour doughs. p. L-1 in Proceedings of the International Symposium on advances in Baking Science and Technology. Kansas State University, Manhattan, Kansas. September 27-28.

Ferry, J. D. 1970. Viscoelastic Properties of Polymers. J. Wiley and Sons. New York, NY.

Hamann, D. D. 1969. Dynamic mechanical properties of apple fruit flesh. Trans. ASAE. 12:170.

Heaps, P. W., Webb, T., Eggett, P. W. P. and Coppock, J. B. M. 1967. Studies on mechanical factors affecting dough development. J. Food Tech. 2:37.

Hibberd, G. E. and Wallace, W. J. 1966. Dynamic viscoelastic behavior of wheat flour doughs. I.

Linear aspects. Rheol. Acta. 5:193.
Hibberd, G. E. 1970a. Dynamic viscoelastic behavior of wheat flour doughs. II. Effects of water content in the linear region. Rheol. Acta. 9:497.
Hibberd, G. E. 1970b. Dynamic viscoelastic behavior of wheat flour doughs. III. The influence of the starch granules. Rheol. Acta. 9:501.
Hibberd, G. E. and Parker, N. S. 1975a. Measurements of the fundamental rheological properties of wheat flour doughs. Cereal Chem. 52:1r.
Hibberd, G. E. and Parker, N. S. 1975b. Dynamic viscoelastic behavior of wheat flour doughs. IV. Nonlinear behavior. Rheol. Acta. 14:151.
LeGrys, G. A., Booth, M. R. and Al-Baghdadi, S. M. 1981. In: Cereals A Renewable Resource. Eds. Y. Pomeranz and L. Munck. American Association of Cereal Chemists. St. Paul, MN.
Navickis, L. L., Anderson, R. A. Bagley, E. B. and Jasberg, B. K. 1982. Viscoelastic properties of wheat flour doughs: Variation of dynamic moduli with water and protein content. J. Texture Studies. 13:249.
Rao, V. N. M. 1984. Dynamic force-deformation properties of foods. Food Tech. 38:104.
Matsumoto, S. 1979. Rheological properties of synthetic flour doughs. pp. 291-302. In: Food Texture and Rheology. Ed. P. Sherman. Academic Press. London.
Smith, J. R., Smith, T. L. and Tschoegl, N. W. 1970. Rheological properties of wheat doughs. III. Dynamic shear modulus and its dependence on amplitude, frequency and dough composition. Rheol. Acta. 9:239.
Schofield, J. D., Bottomley, R. C., Legrys, G. A., Timms, M. F. and Booth, M. R. 1985. Effects of heat on wheat gluten. pp. 81-90. In: Proceedings of the Second International Workshop on Gluten Proteins. Eds. A. Graveland and J. Moonen.
Szczesniak, A. A., Loh, J. and Manell, W. R. 1983.

Effect of moisture transfer on dynamic viscoelastic properties of wheat flour/water systems. J. Rheol. 27:537.

Whorlow, R. W. 1980. Rheological Techniques. Ellis Horwood Ltd. Chicester, UK.

THE USE OF A PENETROMETER TO MEASURE THE CONSISTENCY OF SHORT DOUGHS

Andrew R. Miller
Flour Milling and Baking Research Association
Chorleywood, Herts, ENGLAND

INTRODUCTION

The Biscuit Section at FMBRA commenced a study of manufacturing processes for rotary moulded short-dough biscuits by carrying out a survey of the production problems encountered by the UK biscuit industry. Dough consistency variability and measurement featured prominently. Variability in dough consistency was considered to be a major source of problems in controlling biscuit weight and dimensions. Although the consistency of short doughs was routinely assessed on a subjective basis by bakery mixer operatives there was no generally accepted method of objective measurement, and it became one of the first aims of our studies to develop such a method. It was recognised that the method should be simple in procedure and utilise portable, robust, and inexpensive equipment.

A review of the literature did not reveal an existing method of measuring the consistency of short doughs. Indeed, only one reference (Gorazdovski, 1952) to the rheology of short-biscuit dough was discovered and this was considered an unsatisfactory paper by the author (Muller, 1975) of a review in which reference was made. However, the methods of subjective assessment of short dough consistency were common to many bakeries. The mixer man would compress a sample of dough in his palm and then insert his thumb into the dough ball and on this basis proclaim the dough to be too soft, acceptable or too tough. A second less common test involved pulling the dough ball apart and although many mixer men could not explain why they did this or what they were looking for, this did appear to be a test for extensibility related to gluten development. As extensibility is neither a common nor prominent characteristic of rotary moulded doughs our attention focussed on the thumb test which could be simulated with a penetrometer. In initial studies (Steele, 1977) other instruments including an extruder and a rotating vane shear-tester were evaluated and rejected in favour of the penetrometer. Several penetrometers were used in our studies including the Instron Universal Tester and a hand-held penetrometer used for soil testing. However, when it became available, the Stevens LFRA Texture Analyser, marketed in the USA as the Voland-Stevens Texture Analyser, came closest to meeting our requirements and all the data in this talk are presented in the scale of measurement of this instrument.

DOUGH AND BISCUIT PRODUCTION

The Lincoln biscuit formulation shown in Table 1 is typical of rotary moulded short-dough biscuits produced in the UK. Doughs were mixed in a pilot-scale high speed mixer by a creaming procedure in which all ingredients except flour were creamed for 3min at 75 rpm. After the addition of the flour the dough was formed using 1min mixing at the same speed. The dough

TABLE 1
Lincoln test recipe

Ingredient	% on flour wt
Flour	100
Fat	32.1
Sugar (pulverized)	29.5
Skimmed milk powder	1.79
Salt	1.07
Sodium bicarbonate	0.357
Ammonium bicarbonate	0.179
Water	10.1-14.3

was allowed 20min standing time and the dough pieces were formed in a pilot-scale rotary moulder. Biscuits were baked on steel trays in a pilot-scale forced convection travelling oven. The biscuits were cooled for 20min at ambient temperature before sampling and measurement.

DOUGH CONSISTENCY MEASUREMENT

The method of preparing dough samples was designed as a simple means of removing randomly distributed pockets of air while minimising handling and work softening of the dough. The equipment used is shown in Fig. 1.

Ten minutes after mixing was complete, 105g of dough was weighed into the plastic pot and compressed with the lid patterned with the array of spikes and holes. While the dough was compressed gently, the lid was moved back and forth with respect to the pot and the dough was then compressed to the full extent determined by the lip of the lid. This was repeated until all the pockets of air which could be seen initially between the dough and side wall of the pot had been expelled. At this point it was considered that the body of the dough was also free of air pockets. The first lid was removed leaving an uneven surface as shown which was flattened by pressing the smooth surfaced lid as far as it would go into the pot.

The surface area of the prepared sample was sufficient to permit three consistency measurements

Fig.1. Sample preparation equipment, showing a compressed dough prior to flattening of the sample surface.

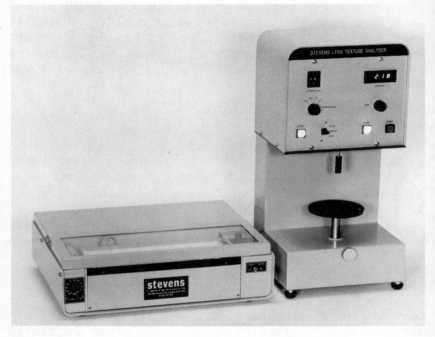

Fig.2. Stevens-LFRA Texture Analyser

per pot. The dough consistency is obtained as the average of six measurements using two pots of dough.

The Stevens-LFRA Texture Analyser (Fig. 2) utilises a load cell to measure the force required to maintain a constant rate of penetration to a preset depth. Our instrument has a 1kg load cell. The desired range of dough consistencies was recorded using a 6.3mm diameter cylindrical head, which was supplied with the instrument, and penetration at 1mm/s to a depth of 15mm. Later models having a 2kg load cell would have permitted the use of a larger diameter head.

The described procedure was found to be operator dependent. The source of this dependency was the dough sample preparation. Table 2 shows an extreme case where one instrument operator obtained significantly different measurements from samples of the same dough prepared by different people. Samples prepared by person 2 gave the highest measurements for doughs which covered a wide range of consistency. All the results discussed in the remainder of this talk were obtained using a single operator to carry out the whole measurement process.

For a single operator the reproducibility of consistency measurements of 10 doughs of each of three levels of consistency are shown in Table 3. The consistency levels were obtained by varying the recipe water level. The coefficients of variation are very similar for a wide range of consistency measurements.

CORRELATION WITH SUBJECTIVE ASSESSMENTS

Over a three-year period we recorded penetrometer measurements and subjective assessments of softness of doughs containing different biscuit flours and recipe water levels. The next two graphs summarise our results but for clarity the data points have been

TABLE 2
Source of operator dependance of dough consistency measurements

Dough type	Person who prepared the sample	Dough consistency	
		mean g	standard error of the mean g
Soft	1	144	4.2
	2	184	5.3
Tough	1	288	4.9
	2	348	5.9

TABLE 3
Reproducibility of dough consistency measurements

Dough type	Dough consistency	
	mean g	CV %
Soft	161	8
Average	238	5
Tough	326	5

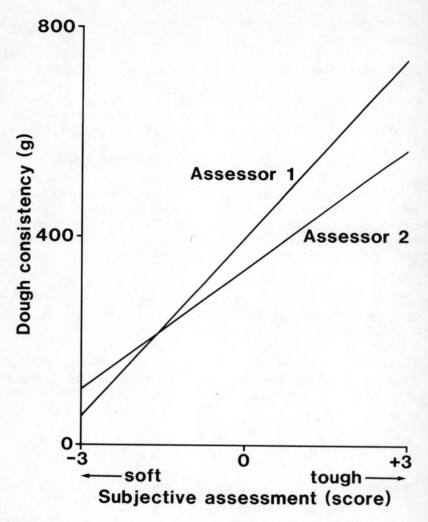

Fig.3. The relationships between dough consistency measurements and subjective scores recorded by two assessors.

removed. The first of these (Fig. 3) shows that two assessors can measure consistency on two completely different scales. In both instances the relationships were reasonably correlated with penetrometer measurements, the correlation coefficients being 0.83 and 0.85 for assessors 1 and 2 respectively. Over a period, the assessments of a single person are not

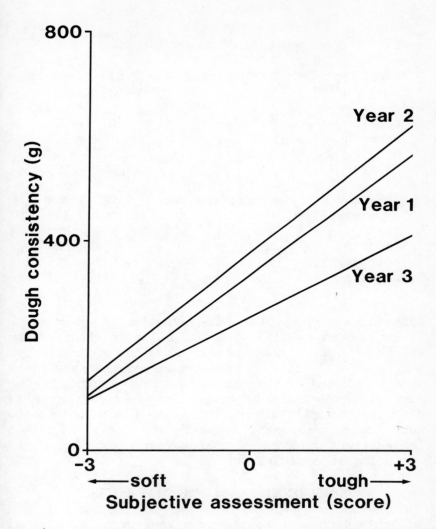

Fig. 4. The relationships between dough consistency measurements and subjective scores recorded over a three year period by a single assessor.

necessarily consistent (Fig. 4). Assessments recorded in experiments in years 1 and 2 were significantly displaced and in year 3 were given scores on a significantly different scale. For the three experiments as a whole a superior correlation of dough piece weight with penetrometer measurements confirmed that it is the subjective assessments of consistency

TABLE 4
Measurements of production doughs

Dough Type	Machining	Range of dough Consistency, g
Cookie	Sheeted	155 - 284
Bourbon	↓	284 - 328
Nice		284 - 355
Ginger		399 - 483
Shortcake 1	Rotary moulded	125 - 260
Shortcake 2	↓	192 - 243
Pastry case		196 - 321
Lincoln		209 - 226
Shortcake 3		243 - 395
Shortcake 4		311 - 361

which are most variable. This variation in assessments is attributed to the absence of reliable standards to which to relate doughs being tested subjectively. Accurate subjective assessment necessitates a good memory and some experience in handling doughs. During a single experiment of several days' duration it is difficult to maintain a single scale of assessment, but over several experiments carried out intermittently over a three-year period it is virtually impossible.

MEASUREMENTS OF COMMERCIAL DOUGHS

Table 4 shows consistency measurements of a range of both sheeted and rotary moulded short doughs produced in the UK. For rotary moulded doughs which have stood for 10min a representative average consistency measurement is 234g.

PREDICTION OF DOUGH WATER ABSORPTION

Over a three-year period we tested 18 commercially available biscuit flours at six different recipe water levels. The range of relationships between dough consistency measurements and recipe water level is

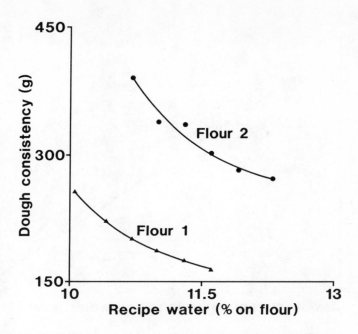

Fig.5. Relationship between dough consistency and recipe water level for two different flours.

illustrated graphically for the two extreme flours in Fig. 5. Linear relationships are obtained by using the base 10 logarithm of the consistency measurements. The relationships differ from one flour to another.

The target dough consistency of our pilot-scale test procedure for rotary moulded short-dough biscuits is 234g. This is the average of the measurements of commercial doughs previously referred to. The optimum recipe water level for a particular flour is therefore defined as that which produces a dough consistency of 234g. The experimental analysis was reconsidered in terms of the percentage deviation of the test recipe water level from the optimum level. This analysis generated a standard relationship between dough consistency and recipe water level which was applicable to all the 18 flours tested. This relationship is illustrated in Fig. 6. Equation 1 describing this relationship is shown below. In this equation, W_t is the test recipe water level, which produced a dough of consistency C g and W_o is the

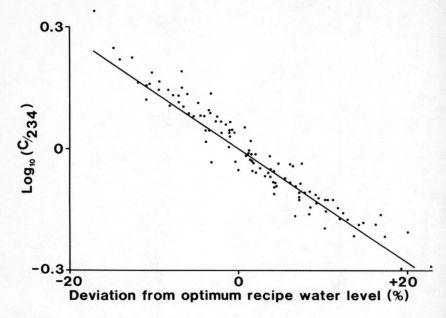

Fig.6. Relationship between $\log_{10}(C/234)$ and the deviation from optimum water level.

optimum recipe water level which would produce a dough consistency of 234g.

$$100 \frac{(W_t - W_o)}{W_o} = -71 \log_{10}(C/234) \qquad (1)$$

Rearrangement of the top equation yields a form in which the optimum recipe water level (W_o) for a test flour can be determined from the dough consistency measurement of a single dough containing the test flour, provided that the test recipe water level is known.

$$W_o = \frac{W_t}{1 - 0.71 \log_{10}(C/234)} \qquad (2)$$

This relationship has been tested using 10 commercial biscuit flours which were not used to derive the equation. The experimentally determined and calculated optimum recipe water levels are shown in

TABLE 5
'Optimium' recipe water levels (% on flour) of ten commercially milled biscuit flours

W_o	W_o^*	$W_o^* - W_o$
11.2	11.9	0.7
11.5	11.6	0.1
11.6	12.0	0.4
11.6	12.2	0.6
11.6	11.9	0.3
11.7	11.5	-0.2
11.9	12.7	0.8
12.1	11.8	-0.3
12.1	12.4	0.3
12.6	12.4	-0.2

W_o is experimental W_o^* is calculated

Table 5. Consideration of the data used to derive the equation suggests that the error in predicted recipe water level would be of the order of ± 0.6% on a flour basis. This corresponds to a range of penetrometer readings of 196-283g on a target of 234g. Under better conditions with the test and optimum recipe water levels reasonably close together, the error of prediction should be smaller than this. With the present penetrometer method the reproducibility of measurements accounts for 40% of this range.

Obviously the derived numerical relationship is not directly translatable into commercial practice. However, the principles on which it has been based could be used by others to derive equations relevant to their own dough formulations.

The ultimate test of any dough measurement technique is whether it can be used to predict the performance of the dough on the production line. In our experience doughs of a wide range of consistency can be rotary moulded. This means that although the penetrometer method could be used to determine whether a dough should be fed to the rotary moulder or rejected, the correct decision can readily be made from subjective assessment of the dough. However, for control purposes where more critical assessment of the

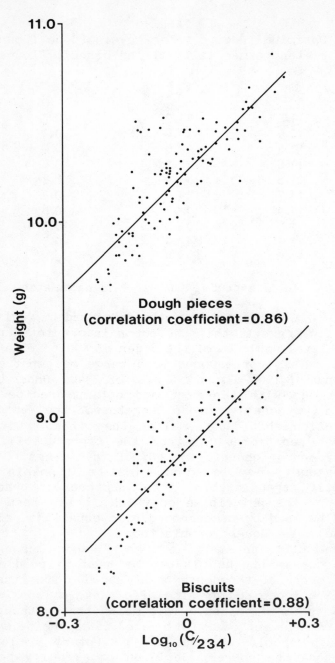

Fig.7. Relationship of dough pieces and biscuit weights with $\log_{10}(C/234)$.

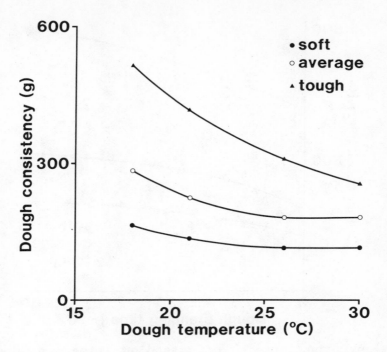

Fig.8. The effect of temperature on dough consistency.

dough is required the penetrometer technique is of potential value as shown in Fig. 7. Under standardised test conditions, dough consistency measurements are well correlated with both dough piece and biscuit weight. Dough consistency was also reasonably correlated with biscuit thickness. Having shown the potential value of our method of dough consistency measurement it is necessary to say that there would be several problems in using the technique in a production control system. Dough consistency measurements are sensitive to production variables other than the properties of the flour used. The sensitivity to dough temperature is shown in Fig. 8. The tougher doughs which were produced using a lower water level were highly sensitive to temperature whereas the softer doughs made using a higher water level were scarcely affected by temperature changes above 23°C. We believe that the precise nature of these effects is dependent on the properties of the fat used in the recipe. The dough consistency

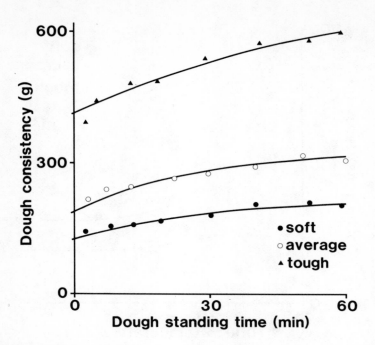

Fig.9. The effect of standing time on dough consistency.

measurements are also sensitive to the standing time of the dough after the completion of mixing. Fig. 9 shows that tougher doughs are more sensitive to toughening with age than are softer doughs. Dough consistency measurements are sensitive to the recipe levels of fat, flour and water as shown in Fig. 10 but are not sensitive to the level of pulverised sugar other than in its role as a diluent of these other ingredients.

The maximum expected metering error of bulk handling equipment used to deliver specified loads of ingredients to the mixer in commercial baking is ±5% of the set batch weight. For this reason the levels of the major ingredients: fat, sugar, flour and water, in a production line dough of specified formulation is liable to slight variation from one mixing to another. Unfortunately these slight variations may cause changes in dough consistency which are detectable by the penetrometer method. Thus an optimum recipe water level determined on the basis of

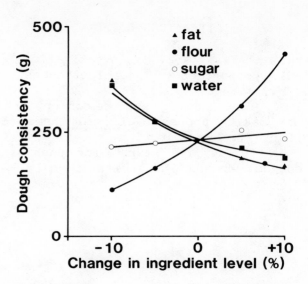

Fig.10. The effect of changes in ingredient level on dough consistency.

measurements of a production line dough may not apply to a subsequent dough because of the background variability in dough composition resulting from metering errors.

CONCLUSIONS

We have developed a method of measuring the consistency of short doughs using a penetrometer. The measurements obtained have practical relevance in that they can be used to determine the recipe water level required to obtain a dough of target consistency from different flours. It is desirable to obtain doughs of target consistency because under standardised test conditions consistency is strongly correlated with biscuit weight and thickness. This method has potential application in problem solving and laboratory matching of production doughs. However, the method would require a full-time operator and is sensitive to operator to operator variability, dough temperature and age and the metered levels of ingredients in the dough. For these reasons, at present, the potential benefits of using this method are unlikely to be fully realised in the production bakery.

LITERATURE CITED

Gorazdovski, T.Ya. 1952. Investigation of the rheological properties of confectioners' dough. Kolloid. Zh., 14:408.

Muller, H.G. 1975. Rheology and the conventional bread and biscuit making process. Cereal Chem., 52, May - June. Part II: 89r - 105r.

Steele, I.W. 1977. The search for consistency in biscuit doughs. Baking Industries Journal, 9, March: 21, 23-24, 33.

GRAIN RESEARCH LABORATORY INSTRUMENTATION FOR STUDYING THE BREADMAKING PROCESS[1]

R.H. Kilborn and K.R. Preston

Canadian Grain Commission
Grain Research Laboratory
1404 - 303 Main Street
Winnipeg, Manitoba
R3C 3G8   Canada

ABSTRACT

Several instruments developed at the Canadian Grain Commission's Grain Research Laboratory for assessing dough and bread properties at different stages of the baking process are described. These instruments involve the use of force transducers, power and energy meters and displacement devices. Concepts and some details of each design are illustrated and discussed. Applications include monitoring of dough mixing properties; dough expansion (height) during fermentation and proofing; dough response to sheeting; dough bulk at time of molding; changes in loaf height during baking and bread crumb properties.

INTRODUCTION

The effect of ingredients and processing conditions upon bread quality has been the subject of many studies. These studies have led to a much better understanding of the relationships of these

---

[1] Paper No. M142 of the Canadian Grain Commission, Grain Research Laboratory.

factors to end product quality in terms of loaf volume, crumb and crust characteristics. However, less information is available concerning physical dough properties during bread processing, and their interrelationships with ingredients, processing conditions and end product quality. This lack of information has, in part, been due to a lack of instrumentation capable of measuring various physical dough parameters during bread processing.

At the Grain Research Laboratory we have been trying to introduce objective, as opposed to tactile, measurements through the use of instrumentation at critical stages of the baking process. Long term objectives of these studies include:

(1) decreasing testing requirements by increasing the number of objective measurements available;

(2) obtaining a better basic understanding of the relationship of physical dough properties to bread quality in terms of ingredients and processing conditions;

(3) use of these measurements under both laboratory and commerical conditions for quality control purposes.

In the present paper, some of the instrumental methods we are presently using will be discussed including mixing energy measurements, dough expansion properties, sheeting and molding properties, loaf expansion characteristics during baking and crumb compression characteristics.

MIXING ENERGY REQUIREMENTS

The first major processing step in the production of bread usually takes place at the mixer. Mixing must not only distribute the dough ingredients to provide a homogenous mass, but for many processes, must work (develop) the dough in such a way that the dough is capable of forming thin sheets (membranes) having the appropriate balance between extensibility and resistance to extension needed for expansion and gas retention. In some processes, the dough structure from the mixer directly determines the final bread structure (other factors being equal). Examples of this include Brazilian french bread, some continuous type breads, and Chorlywood process breads where minimum fermentation and little or no mechanical work is imparted to the dough after mixing. In other processes, the dough structure from the mixer has a less direct effect upon the final bread structure. Examples of this include traditional french bread and some types of straight dough processes where fermentation and mechanical work imparted after mixing is mainly responsible for dough development.

In processes where dough is developed in the mixer, it becomes progressively stiffer until a stage of minimum mobility occurs. If the mixing action is sufficiently intense, the dough will then progressively soften as mixing continues. The three stages described are often referred to as development, peak consistency and breakdown respectively. In most cases these stages can be monitored using a recording dynamometer such as the farinograph or some other torque indicating device such as a force transducer configured to the drive shaft of the mixer or the motor. However, this approach requires considerable mechanical modification to most dough mixers not designed for this purpose and in some instances is not practical. A practical alternative is to monitor the power used by the mixer motor during the mixing operation.

Description of GRL Energy Input Meters

The GRL Energy Input Meter (Fig. 1) is intended for research work where a number of variables have to be applied to obtain power and energy measurements on a weighted and/or otherwise qualified basis (Kilborn, 1979). It may be used to measure power and energy for any 100-130 volt a.c. device having a power consumption of 2,000 watts or less, but is primarily intended for laboratory dough mixers. A d.c. signal ouput which is directly proportional to power may be used for a recorder or other purposes. Any portion of the 2,000 watt signal may be zeroed out and only that portion which exceeds the null

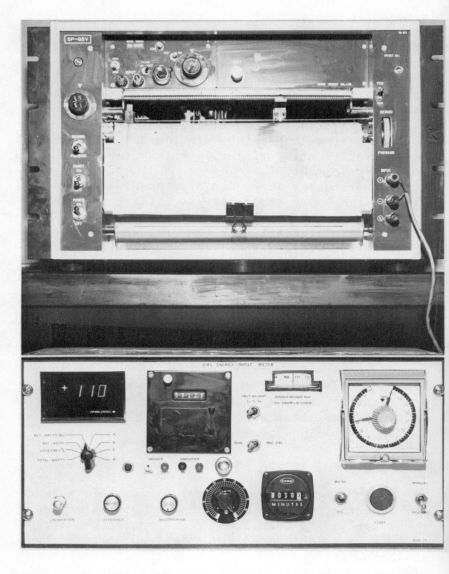

Fig. 1. GRL Energy Input Meter with chart recorder showing mixing curve.

point is measured. This signal may be weighted according to mixer efficiency and dough mass to represent net watts per kg of dough. The energy consumed is accumulated through an integrator by a digital counter on the panel. The counter may be preset to switch off the mixer at a predetermined value. The timer on the panel may alternatively be used to switch off the mixer at a preset time interval. A digital panel meter provides a means of obtaining total watts and net watts per kilogram of dough.

Examples of Usage

Under standard conditions of formula and procedure, different varieties of wheat can produce quite differently shaped mixing curves. Information such as work requirements and mixing times can be determined for peak consistency and the relative mixing requirements of varieties determined under actual baking formulation without carrying out separate mixing experiments.

Mixing curves are useful in routine tests and reduce the number of bakes required. There is increased confidence in baking results where replicate bakes are done on separate days, particularly if a fermentation period has preceeded the final mix such as in the Remix baking procedure. Matching curves verify that the formulation and fermentation stages were similar. Further, the exploratory work required to satisfy a different mixing requirement is reduced to a minimum.

In the example shown in Fig. 2, the bread on the left was obtained from a medium strong flour. The top loaf is from 100% test flour using the Remix baking test (constant 2.5 min remix time). The bottom loaf is a blend of 50% test flour and 50% low protein soft wheat flour. In both cases the mixing requirements were satisfied as is indicated by the mixing curves.

The centre row of bread was obtained from a very strong flour. The 100% test flour loaf was very poor. It can also be seen from the mixing curves that the mixing peak for the 100% very strong wheat flour was not reached. However the Remix blend method produced a satisfactory loaf of bread. This happens because the addition of the low protein

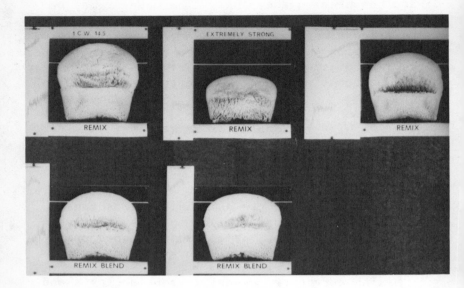

Fig. 2. Remix and Remix blend mixing curves for bread from 1 CWRS (left) and 1 CU (middle) wheat flour. Right shows Remix-to-peak for 1 CU.

flour lowers both the mixer critical speed requirements and the minimum energy requirement through the dilution of the test flour.

It remains then to do one more test mixing the 100% strong wheat flour to peak consistency. If the loaf volume is in line with that projected from the blend test, then we can conclude that the flour has a high mixing energy requirement. If the loaf volume is significantly lower than the projected value, then we conclude that the flour is one having very high critical mixing speed characteristics (ie. mixer speed must be increased to allow development of the dough).

Normal baking ingredients other than flour can have a significant effect on the curve shape, work, level and mixing time. For example, as salt is reduced, work and mixing times shorten. This effect is more pronounced with some flours. As earlier papers have shown, fermentation also shortens the mixing curve and reduces mixing requirements (Preston and Kilborn, 1982).

Energy values obtained with energy mixing input meters can complement mixing time in evaluating the mixing response or tolerance of flours to different mixer types and mixing conditions. This is due, in part, to the fact that mixing time is not linearly related to mixing energy input. This is demonstrated in Fig. 3 which shows the mixing curve (GRL mixer) of a very strong wheat flour (No. 1 Canada Utility) and the bread produced at 50% peak time, 50% peak work and at peak using the Canadian Short Process baking procedure. As can be seen, the quality of bread produced from mixing to 50% peak time is much different (inferior) to that produced from mixing to 50% peak work or energy.

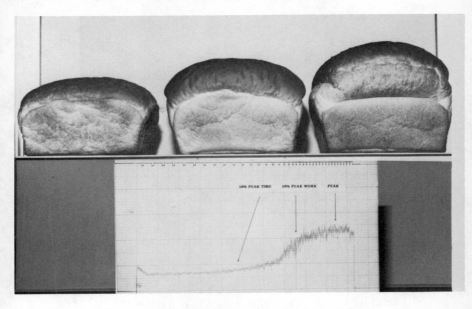

Fig. 3. Canadian Short Process bread prepared from No. 1 CU flour; doughs mixed to 50% peak time, 50% peak energy and to peak.

The principle of monitoring dough development has been applied in some commercial conditions where the mixing action is sufficiently intense.

This can provide a means for the baker to easily determine when doughs are sufficiently developed. Fig. 4 shows a curve obtained from a Kemper 15 spiral mixer using 5 kg of dough. In this case, the peak development of the dough is clearly indicated from the curve. Fig. 4 also shows a curve obtained from a Canadian plant bakery employing the power measuring system to monitor their production doughs from a 2,000 lb bar mixer.

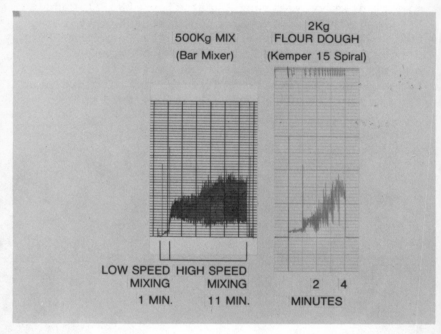

Fig. 4. Mixing curves obtained from bar mixer and spiral mixer.

The Tweedy mixers seem to be an exception. While they are capable of rapid intense mixing action, the development of dough is not clearly reflected in the power consumption of the mixer motor. A special probe involving a force transducer attached directly to the mixing bowl is required to monitor dough development (Kilborn and Tipples, 1981).

DOUGH HEIGHT CHARACTERISTICS

Changes occurring in the gas retention properties of doughs during fermentation and proofing play an important role in end-product quality. Several instruments have been described to assess these properties either in terms of dough volume (Bloksma, 1971) or dough height (Shuey, 1975; Marek and Bushuk, 1967; Garnetz et al, 1949). However, with most of these instruments, dough being studied cannot be further processed into bread.

The GRL dough height tracker can be used to follow the expansion (or fall) of doughs during fermentation and proofing (Kilborn and Preston, 1981). It consists of a light weight low friction piston that is placed on the dough surface. The movement of the piston is translated into an electrical signal proportional to the dough height. The instrument is designed for small to medium scale doughs (300 g of flour or less) and can measure changes in dough height of up to 100 mm. No noticeable dough deformation is produced by the instrument, allowing further processing without any adverse effects. Cylindrical containers of uniform size are used for fermenting sponges and/or doughs. In measuring proofing performance of doughs, panned doughs give the most reliable results owing to the dough being largely restricted on all sides. The expansion of hearth type breads is not totally reflected in height, and therefore using the height measurement as an index of expansion is complicated by the flow of the dough. However, the height may be an indicator of the boldness of the dough.

The tracking may be used to compare the performance of different flours. In the example of sponge doughs (Fig. 5), a red spring flour is compared with a red winter flour. Differences in the fermentation response of these samples are obvious.

Figures 6 and 7 show curves for the dough recovery after mixing and proof stages, respectively, for the two flours. In the example shown in Fig. 6, the normal dough recovery time of 30 min was extended to 100 min with no punching. Differences

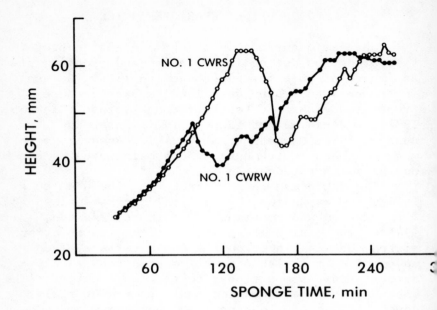

Fig. 5. Sponge curves for CWRS and CWRW wheat flours.

in the initial (zero time) heights of the two doughs did not change up to about 40 min of dough time, but thereafter the height of the CWRS dough increased more rapidly than did that of the CWRW dough. This suggests that the gas retention properties of the CWRS dough were greater than those of the CWRW dough. Differences in proof heights between the two doughs (Fig. 7) were even more dramatic. Differences in dough heights were evident after only 10 min. After the normal proof time of 70 min, the CWRS dough was 21 mm higher than the CWRW dough. This difference in gas retention properties is probably partially responsible for the larger loaf volumes of bread produced from CWRS wheat flours.

The tracker may also be used to evaluate the effects of ingredients other than flour. For example, we routinely use sponge curves to determine yeast quality. Yeast shipments of inferior quality, which are not necessarily reflected in lower gassing

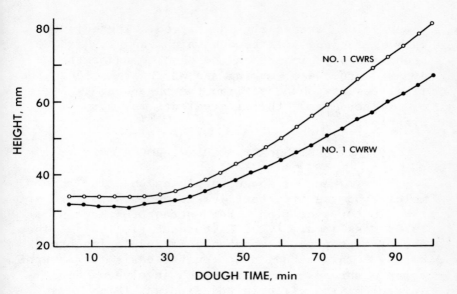

Fig. 6. Dough curves after mixing for CWRS and CWRW wheat flours.

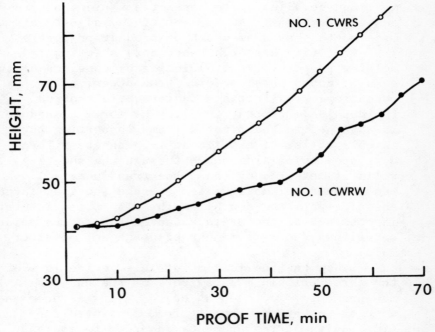

Fig. 7. Proof curves after mixing for CWRS and CWRW wheat flours.

power value, are easily identified. Ingredients such as fat and milk powder which can vary from brand and/or shipment can be checked under the dynamic process of baking and may be related to the final results obtained from the same dough. Effect of oxidation and other ingredients may also be clearly seen.

SHEETING AND MOLDING PROPERTIES

The makeup of dough involving sheeting and molding has an important influence on bread properties in terms of final crumb structure (Kamman, 1970; Moss et al, 1979; Kilborn et al, 1981). The sheeting step is also capable of developing bread doughs. This method has been the basis for several types of bread processing which involve minimal initial mixing (Kilborn and Tipples, 1974; Moss, 1980; Kilborn et al, 1981). In most commercial baking situations, dough sheeting and molding equipment can be adjusted to correct a variety of processing problems. At present, these adjustments are made on the basis of dough or product properties.

The Grain Research Laboratory sheeting and molding property (Fig. 8) indicator consists of two sensing components, a force transducer, and a potentiometer with the associated electronics (Kilborn and Preston, 1982). The force transducer is attached to the pivoted frame supporting the movable roller of a National 1-lb sheeter. Force developed by passing dough through the sheeting rolls is measured by the force transducer, allowing measurements of maximum and average sheeting force, sheeting work and dough length. The potentiometer is attached to the Grain Research Laboratory molder and allows the measurement of dough volume after molding.

Since the publication of this device in 1982, the three principle parameters of the sheeting measurements, i.e. dough length, average force and work are obtained directly from a printed tape in final units of measurements in order to increase efficiency by eliminating the rather long time

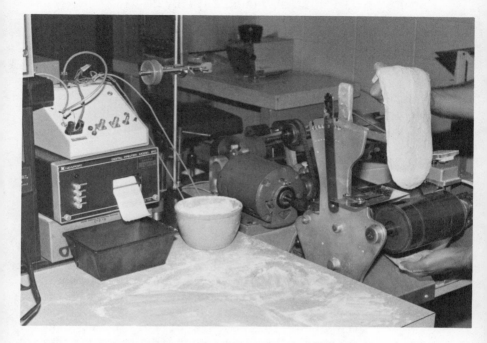

Fig. 8. GRL Sheeting and Molding Property Indicator.

needed to make the post measurements (which include area) and calculations. This was done through electronic integration and analogue to digital techniques.

Most of our baking processes involve three sheetings at make up. We are currently ignoring the first sheeting in making assessments - considering the first sheeting as a "shaping of dough piece" operation.

The length of the dough piece divided by the work or energy value, appears to parallel the bakers sensory evaluation over the range of from very extensible to very bucky. While these extremes can be assessed easily without instrumentation it is hoped that more subtle differences can be quantified. On line bakery control of sheeting pressure is a possible future use of this type of instrument.

LOAF TRACKER IN OVEN

A device somewhat similar to the dough height tracker has been installed in special ovens employing the heat sink oven principle. This total package will be described in a subsequent publication. This instrument makes it feasible to track the changes in the height of the loaf that occur during baking.

CRUMB COMPRESSION CHARACTERISTICS

The Grain Research Laboratory compression tester is a versatile, relatively inexpensive instrument for the measurement of the textural properties of bread crumb (Kilborn et al, 1983). A plunger attached to a force transducer is moved vertically by a variable-speed motor-driven cam that provides precise, repetitive compression of foods either to constant deformation or to constant force in conjunction with a comparator.

As with other instruments of this type, the compression tester has been used for the measurement of the crumb compression characteristics of various types of bread after baking to determine textural properties and can be used to assess the effect of ingredients and processing conditions on these properties in terms of firmness, stress relaxation, cohesiveness and stickiness.

Table 1 shows the effects of bread storage time on the textural properties of Chorleywood processed bread using two types of flours. As expected, firmness values for the bread increased during storage. In addition, stress relaxation, which is a measure of springiness, increased and cohesiveness decreased. The higher volume, higher absorption bread produced from the CWRS wheat flour generally gave lower, more desirable firmness values than the CWRS-CEWW blend.

SUMMARY

At the Grain Research Laboratory, instrumental

TABLE 1

Effect of Flour Type and Loaf Storage Time on the Textural Properties of GRL Chorleywood-Processed Bread

Flour Blend	Storage Time (days)	Firmness (N/m)	Stress Relaxation (%)	Cohesiveness
100% CWRS	0.75	4,360	16.8	0.890
	1.75	6,310	18.5	0.858
	2.75	6,830	20.3	0.854
	3.75	8,480	22.7	0.827
	6.75	11,180	27.0	0.772
55% CWRS/ 45% CEWW	0.75	5,360	19.7	0.866
	1.75	7,190	24.2	0.817
	2.75	8,580	26.8	0.783
	3.75	9,080	27.7	0.769
	6.75	11,030	33.1	0.743

methods have been developed to assess the physical properties of doughs during the various stages of bread production. Future use of these types of instrumentation will provide both the researcher and commercial baker with improved techniques for studying and controlling the breadmaking process.

REFERENCES

Bloksma, A.H. 1971. Rheology and chemistry of dough. Page 523 in: Y. Pomeranz, ed. Wheat: Chemistry and Technology. Am. Assoc. Cereal Chem.: St. Paul, MN.

Garnatz, G., Hodler, P.W. and Rohrbaugh, W. 1949. An evaluation of cabinet fermentation with commercial size sponges and doughs. Cereal Chem. 26:201.

Kamman, P.W. 1970. Factors affecting the grain and texture of white bread. Bakers Digest 44(2):34.

Kilborn, R.H. 1979. Mixing curves and energy measurements in test baking. Bakers Journal Oct/Nov:18.

Kilborn, R.H. and Preston, K.R. 1981. A dough height tracker and its potential application to the study of dough characteristics. Cereal Chem. 58:198.

Kilborn, R.H. and Preston, K.R. 1982. A dough sheeting and molding property indicator. Cereal Chem. 59:171.

Kilborn, R.H. and Tipples, K.H. 1974. Implications of the mechanical development of bread dough by means of sheeting rolls. Cereal Chem. 51:648.

Kilborn, R.H. and Tipples, K.H. 1981. A novel device for sensing changes in dough consistency during dough mixing I Application with Tweedy mixer. Bakers Journal April/May:16.

Kilborn, R.H., Tipples, K.H. and Preston, K.R. 1983. Grain Research Laboratory compression tester: Its description and application to measurement of bread-crumb properties. Cereal Chem. 60:134.

Kilborn, R.H., Tweed, A.R. and Tipples, K.H. 1981. Dough development and baking studies using the pilot scale dough broke. Bakers Digest 55(1):18.

Marek, R.H. and Bushuk, W. 1967. Study of gas production and retention in doughs with a modified Brabender oven-rise recorder. Cereal Chem. 44:300.

Moss, H.J. 1980. Strength requirements of doughs destined for repeated sheetings compared with those of normal doughs. Cereal Chem. 57:195.

Preston, K.R. and Kilborn, R.H. 1982. Sponge-and-dough bread: Effects of fermentation time, bromate and sponge salt upon the baking and physical dough properties of a Canadian Red Spring Wheat. J. Food Sci. 47:1143.

Shuey, W.C. 1975. Practical instruments for rheological measurements of wheat products. Cereal Chem. 52:421.

LABORATORY MEASUREMENT OF DOUGH DEVELOPMENT

P.J. Frazier, C.S. Fitchett and P.W. Russell Eggitt
Dalgety U.K. Limited, Group Research Laboratory,
Station Road, Cambridge, CB1 2JN, U.K.

INTRODUCTION

A variety of commercially available recording mixers have been used over many years for the dynamic monitoring of dough development. These allow relative comparisons of flour quality but are generally unsuitable for studying flour performance in short-time breadmaking processes involving high-energy mixing, such as the Chorleywood Bread Process, where total energy input to the dough is carefully controlled and the rate of work input is also an important factor. When this subject was last reviewed (Frazier et al., 1975; Daniels & Frazier, 1978), it was found that, although recording mixers had been used almost exclusively to predict optimum development, tests on rested doughs (which represent much more closely the dough condition on entering the baking oven after proof) could lead to very different conclusions regarding optimum development.

Notwithstanding the obvious difference in instrumental technique, one factor suggested as contributing to the discrepancy between peak development predicted from dynamic mixing and from tests on rested doughs was a possible increase in free water content with mixing which could lead to a premature reduction in torque. However, it is questionable whether such a change would not also be seen in the rested system. Direct measurement of freezable (free) water by differential scanning calorimetry has previously been carried out on mechanically developed doughs, but only at a single level of work input (Davies & Webb, 1969). Results for a wide range of work inputs are presented in this present paper.

An alternative explanation is that the peak discrepancy could be a reflection of the time dependence of the dough system. That wheat flour dough is a rheologically time-dependent material was appreciated in probably the earliest type of dough test, described by Halliwell in 1898, in which flour properties were initially assessed by hand-working flour and water into a dough and then reassessed about 30 min later by stretching the dough manually. Mechanical versions of these two procedures have now been in use for many years and the designers of two typical machines (the Brabender Farinograph Mixer and the Extensograph) originally pointed out that doughs undergoing mixing (in an 'excited condition') exhibited different physical properties from those shown in a state of relaxation (Munz & Brabender, 1940a,b).

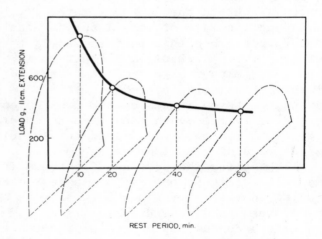

Fig. 1. Construction of a structural relaxation curve from dough extensograph resistance at a fixed extension (in this example, 11cm) after different rest periods (from Dempster, et al., 1953).

Hlynka and co-workers (Dempster et al., 1952, 1953, 1955) examined in detail the changes undergone by doughs when left for increasing times after mixing and moulding prior to testing. The decrease in resistance to extension which they found to occur with time, was called 'structural relaxation' (Fig. 1) and could be described by an exponential or simpler hyperbolic equation. However, this behaviour was only documented for conventional, low-work-input doughs. The present paper examines the structural relaxation of mechanically developed doughs and relates this to the interpretation of optimum development.

EXPERIMENTAL

Materials

Two untreated, unbleached, commercially-milled flours were used in this investigation: a strong flour (100% U.S. Northern Springs) and a weak flour (100% English) having moisture contents of 13.7% & 13.6% respectively and, on a 14% moisture basis, protein: 13.3% & 9.5% and water absorption (500 BU): 63.9% & 54.6%. In addition, flour was prepared from a number of single-variety wheat samples by Buhler milling.

Methods

Dough Mixing

Doughs of constant weight (500 g) containing distilled water 9% below Farinograph absorption at 500 BU and 2% salt (flour wt basis) were mixed under an atmosphere of air or nitrogen (Frazier et al., 1973) in a stainless steel-clad Farinograph bowl attached to the Compudomixer (Frazier et al., 1975). A constant rate of work input (20 kJ/kg.min) was used to obtain doughs over a range of work levels from 2 - 400 kJ/kg. The reduction in dough water content was employed to facilitate rheological testing at high work levels.

Determination of Freezable Water

Freezable water in mechanically developed doughs was determined by differential scanning calorimetry using a Dupont 900 Differential Thermal Analyser and following the method of Davies & Webb 1969. Bound water, defined as that water remaining unfreezable at $-50^{\circ}C$, was calculated by difference between the known total water content of the dough and the amount of freezable water indicated by the endotherm peak.

Rheological Tests

   <u>Torque recording</u>. The torque output signal from the Compudomixer was recorded continuously during mixing to give an indication of optimum consistency similar to a conventional Farinogram, though at a much higher and varying speed in accordance with constant power mixing. In addition, further mixes were carried out during which the torque signal was sampled for 30 s at a mixer speed of 20 rev/min after various levels of work input, using a recorder with zero suppression and a higher amplification to obtain an accurate measurement of torque trace amplitude.

   <u>Structural relaxation tests</u>. A procedure basically similar to that of Dempster et al. (1952, 1953, 1955) was followed, the main modification being the use of an Instron Universal Testing Machine (Model 1122) in place of the Extensograph. Dough samples (100 ± 1 g) were weighed as rapidly as possible immediately following the completion of mechanical development and moulded into cylinders of constant cross-section (approx. 150 mm long by 25 mm diameter) using a Mono Universal Table Moulder (sheeting roll gap: 1/8 inch (3.2 mm); moulding pressure: 1 unit). The dough cylinders were pegged into standard Extensograph cradles and stored at $30^{\circ}C$ in a humidity cabinet (95-100% RH) for various rest periods between 1 and 60 min, timed from the completion of mechanical development. (In practice it was necessary for the 1 min dough sample to be tested immediately after moulding, without storage in the humidity cabinet.)

   Stretching of the dough cylinders was carried out on the Instron previously fitted with an Extensograph cradle holder and hook (Fig. 2) and using a crosshead speed of 500 mm/min. Graphs of force vs vertical extension resembled conventional extensograms except that the load axis was rectilinear rather than

Fig. 2. Stretching dough using an Extensograph cradle holder and hook fitted to an Instron.

curvilinear, the full-scale deflection could be set to 5, 10 or 20 N (approx. 400, 800 or 1600 BU) and no cradle depression occurred, so eliminating the need for correction (Dempster et al., 1953; Muller et al., 1961). From each graph the force required to produce a dough extension of 70 mm was measured. (This length was chosen as being the maximum extension before onset of rupture of the least extensible dough.) Structural relaxation curves were then constructed showing the change in load at 70 mm extension with rest period, each experimental point being the mean of 3 - 6 replicate tests on freshly mixed and rested doughs.

Stress relaxation tests. In addition to the structural relaxation studies, dough development profiles were obtained by compressive stress relaxation measurement (Frazier et al., 1973; Heaps et al., 1968). After mixing to the required work input, replicate 10 g samples of dough were moulded into balls, allowed to rest for 45 min in controlled temperature and humidity ($30^{o}C$, 95-100% RH) and compressed on the Instron between parallel plates closing at 10 mm/min up to a preset load of 1.8 N. The internal stress was then allowed to relax at constant deformation and the time for a relaxation of 1.0 N (approx 100 gf) recorded. This provides a sensitive measurement of the state of dough development - the higher the relaxation time, the stronger or more developed the dough protein structure - and the results correlate well with baking performance (Daniels & Frazier, 1978; Frazier et al., 1979; Frazier, 1979).

RESULTS AND DISCUSSION

Bound Water in Mechanically Developed Doughs

Endotherms were run in triplicate on three samples from each dough, providing nine replicates in all. Mean values of bound water for both flour types at five levels of work input are shown in Table I and for both mixer atmospheres at three work levels on the strong flour only are shown in Table II.

It is evident that very little change in bound water content occurred across the wide range of work levels from 10 to 400 kJ/kg. When differences between flour types were eliminated, variation in bound water with work input just reached the 5% (*) level of significance. However, this was due to a slight fall in values at 60 & 100 kJ/kg followed by a rise again, rather than to any progressive decrease in bound water with increasing work. Small differences in bound water between strong and weak flours were seen at each comparable work level and over all work inputs this just reached the 5% significance level. Considering the great difference in protein content and water absorption between the two flours, the variation in bound water content is remarkably small, supporting the conclusion of Davies & Webb (1969) that there was

TABLE I
Effect of Work Input and Flour Strength
on Bound Water in Dough

Flour Type	Mean Bound Water (g % dry matter)					
	Work Input (kJ/kg)					Overall Work ± standard error
	10	60	100	200	400	
Strong (USNS)	31.0	28.9	28.7	31.4	31.0	30.2
Weak (English)	31.8	31.3	30.3	32.0	32.2	31.5
						±0.36 (*)
Overall Type ± standard error	31.4	30.1	29.5	31.7	31.6	Overall Flour Type & Work 30.9 ±0.25
			±0.57 (*)			

TABLE II
Effect of Work Input and Mixer Atmosphere
on Bound Water in Dough from Strong Flour

Mixer Atmosphere	Mean Bound Water (g % dry matter)			
	Work Input (kJ/kg)			Overall Work ± standard error
	10	100	400	
Air	31.0	28.7	31.0	30.2
Nitrogen	30.2	32.5	29.3	30.7
				±0.56 (ns)
Overall Atmos ± Standard error	30.6	30.6	30.2	Overall Atmosphere & Work 30.5 ±0.39
		±0.68 (ns)		

no difference in bound water between strong and weak flours. When mixing was carried out under nitrogen, Table II, a small rise rather than a fall in bound water was observed at 100 kJ/kg, consequently mean values over both atmospheres showed no significant difference with work input. Similarly, on eliminating effects of work, mixer atmosphere was found to have no significant effect on bound water.

In total, the results indicate a bound water content for dough of 30.7 $\pm$0.3 g % dry matter (equivalent to 23.5% moisture) irrespective of flour type, work level or mixer atmosphere. This is in good agreement with the value of 33 $\pm$1.6 g % found by Davies & Webb (1969) for strong and weak doughs mixed at a single work level in air, and a value of 30 g % found by Bushuk & Mehrotra (1977b) for three cultivars of different mixing strength. The latter workers found no effect on bound water of mixing times between 8 & 20 min in a Farinograph bowl (approx. 25-60 kJ/kg) and, taken together with the present results over a very wide range of mechanical work input, it must be concluded that the decrease in mixing consistency beyond peak torque (eg see Fig. 10) is not due to a decrease in bound water content defined as non-freezable at -50°C. Nevertheless, differences in the mobility of the free water phase, not detected by the DSC freezing mode, may influence dough rheological properties and could explain the extrapolated dough extensibility results of Webb et al.(1970) and Daniels (1975). Possible further evidence for this has been provided recently by Bushuk & Mehrotra (1977a) using differential thermal analysis in the boiling mode.

Structural Relaxation Behaviour

Doughs mixed in Air

A series of structural relaxation curves is shown in Fig. 3 for strong flour doughs mixed in air over a wide range of mechanical work input (10 - 400 kJ/kg). At low work levels (up to 40 kJ/kg) dough resistance to extension fell rapidly with time at first, but the rate of change slowly decreased to approximately zero after about 40 min. Dempster et al. (1952) reported

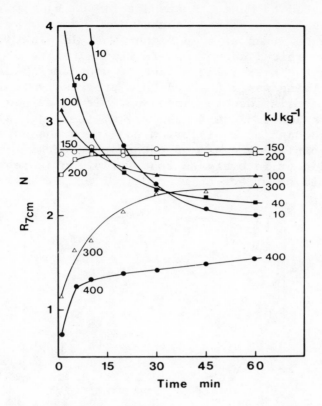

Fig. 3. Structural relaxation curves of resistance at 7 cm extension, derived from Extensograms of doughs developed in air to the work levels indicated.

similar structural relaxation curves for non-bromated, low-work, air-mixed doughs.

However, with increasing levels of mechanical development, the change in dough resistance to extension with time became smaller and the curves flattened out progressively earlier and at higher values of resistance until, at 150 kJ/kg, little or no change occurred with rest period. Further mechanical development then produced a reversal in dough behaviour: thus at 200 kJ/kg, instead of structural relaxation, a small increase in resistance occurred with time ('structural recovery'). Similar behaviour was reported by Munz & Brabender (1940a,b) for 'extensively over-mixed doughs'.

Doughs mixed at even higher work levels (300, 400 kJ/kg) exhibited even greater 'structural recovery' although the overriding effect of dough breakdown due to overmixing was reflected in a progressively lower limiting resistance to extension. It is evident that, at a work level close to 150 kJ/kg, the dough system passed through an equilibrium condition characterised by the complete absence of either structural relaxation or recovery so that variation in rest period had no effect on rheological properties.

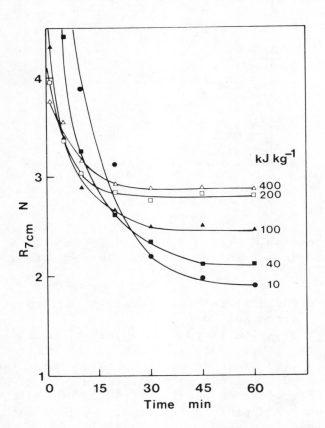

Fig. 4. Structural relaxation curves of resistance at 7cm extension, derived from Extensograms of doughs developed in nitrogen to the work levels indicated.

Doughs Mixed Under Nitrogen

Structural relaxation curves for strong flour doughs mixed under nitrogen over a wide range of work input are shown in Fig. 4. Between 10 & 100 kJ/kg the curves are very similar to those for the equivalent air-mixed doughs (Fig. 3). At high work levels in nitrogen, however, no equilibrium condition or reversal to structural recovery occurred. Instead, increasing work from 100 to 400 kJ/kg resulted in progressively flatter structural relaxation curves and the absence of mechanical breakdown on extensive mixing under nitrogen (Daniels & Frazier, 1978; Frazier et al., 1973) was shown by the absence of any fall in value of the limiting resistance to extension.

Dough Development Assessed after Resting for 45 Minutes or Longer

Doughs Mixed in Air

Dempster et al. (1952) noted that the rheological properties of conventional, low-work doughs had approximately stabilised after 45 min rest. Fig. 3 shows that curves for all work levels, whether exhibiting structural relaxation or structural recovery, had also approximately reached a limiting, stable resistance by 45 min rest. Taking a vertical section through this family of curves at 45 min and plotting resistance against work input gives a development profile with a maximum at 150 kJ/kg (Fig. 5) similar to those obtained previously using stress work or stress relaxation parameters (Frazier et al., 1975).

Thus mechanical development initially strengthens dough, resulting in a progressively greater limiting resistance to extension which, as shown in Fig. 3, is preceeded by a decreasing amount of structural relaxation. A point of maximum development is then obtained at which resistance to extension is greatest and where little or no change occurs with time. Further mechanical work then begins to break down the dough structure, resulting in a progressively smaller limiting resistance to extension which is preceeded by some structural recovery.

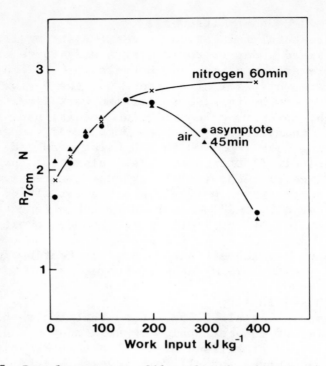

Fig. 5. Development profiles for doughs mixed in air and nitrogen, derived from structural relaxation curves at rest periods of 45 min or longer.

The contribution to the limiting dough strength value, made by the structural relaxation or recovery which precedes stabilisation, may be assessed by treating the curves mathematically as a family of equilateral hyperbolae (Dempster et al., 1955). This enables the asymptotic resistance (ie the resistance at infinite rest time) to be calculated from the equation:-

$$(R - R_a)t = C$$

where R is the resistance at time t, R_a is the asymptotic resistance and C is a constant. Rearranging gives:-

$$Rt = R_a t + C$$

so that plotting Rt against t results in a

straight line with a slope equal to the asymptotic resistance. This value for each work level is also shown in Fig. 5 and generally agrees closely with values observed after 45 min rest. Monitoring dough after a rest period of 45 min can therefore be regarded as measuring the ultimate, stable properties of the protein structure attributable to mechanical development. (At very low work levels (10 kJ/kg) some discrepancy is seen between the 45 min and asymptotic resistance values. Here reference to Fig. 3 shows that resistance was still decreasing slowly at 60 min. However, the value at 45 min can be regarded as a good approximation to the final, stable resistance.)

Doughs Mixed in Nitrogen

As shown in Fig. 4, structural relaxation curves for all work levels in nitrogen had also approximately reached a limiting resistance after 45 min rest. A vertical section through this family of curves gave the development profile also shown in Fig. 5. Calculated asymptotic resistances (not shown) agreed closely with values obtained after 45 min rest, except at very low work levels, as above. Thus, under nitrogen, mechanical development initially increases dough strength, but a plateau is then reached and no breakdown is observed on further mechanical work, in agreement with previous findings (Daniels & Frazier, 1978; Frazier et al., 1973). In the absence of dough breakdown, all time-dependent effects were of a structural relaxation type. No structural recovery was observed.

Dough Development Assessed after Resting for Less than 45 Minutes

As indicated in Fig. 3 & 4, at times shorter than 45 min after mixing, dough properties are changing with time at varying rates depending on work level and atmosphere. A vertical section through the family of curves mixed in air, taken at 20 min, is shown in Fig. 6. The resulting 'development profile' is clearly very different from that shown in Fig. 5 and it could quite reasonably be interpreted that mechanical development initially weakens dough, as shown by decreasing resistance to extension, followed by an increase in strength to a maximum and finally by breakdown and decreasing resistance.

However, such interpretation would be erroneous since the higher initial strength and its apparent decrease with work is an artifact resulting from the incomplete structural relaxation of the lower-work doughs. This curve (Fig. 6) bears a striking resemblance to the results obtained by Launay & Bure (1974a) for the variation of dough modulus of elasticity with work input (see their Fig. 8). Reference to their methods (Launay & Bure, 1974b) reveals that doughs were assessed using a cone and plate rheometer after 20 min rest and so would probably not have been in a stable condition.

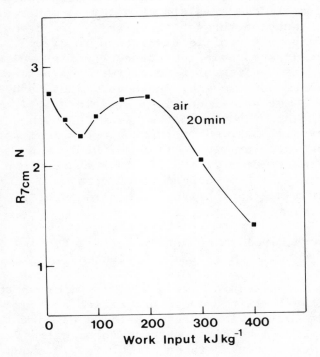

Fig. 6. Apparent development profile for dough mixed in air, derived from structural relaxation curves after a rest period of only 20 min.

It is clearly unwise to observe the rheological behaviour of mechanically developed doughs at any time before structural relaxation or structural recovery has ceased. Observations so recorded can be related only to a changing dough structure at the instant of measurement, with little relevance to dough behaviour in normal commercial bread production using a 40-50 min proof period.

Dough Development Assessed at Zero Rest Time

The use of equipment to record torque at the mixer or power consumption during dough development provides, in effect, an assessment of dough properties at zero rest time, although the method of measurement is necessarily very different from that used in testing rested doughs. By dividing and moulding doughs with the greatest possible speed after mixing and immediately carrying out stretch tests, it was possible to obtain values for resistance to extension after a rest time of only one minute, which could be regarded as a good approximation to 'zero rest time'. Dough properties were then related closely to those present during dynamic mixing, while this method of assessment provided measurements directly comparable with tests on fully rested systems.

The results are illustrated in Fig. 7 and it can be seen that, at rest times approaching zero, dough resistance to extension increased rapidly with work input, reached a peak at about 25 kJ/kg and then fell sharply. Comparison with the mean torque curve, also in Fig. 7, shows good agreement between peak resistance to extension at zero rest and the arrival of the torque curve at its maximum. However, there is little further similarity between the curves, since the torque curve shows far less change with work input beyond the peak.

The comparison is much improved when it is realised that the magnitude of the mixer torque arises largely from the overall resistance to movement of the blades in the dough, upon which are superimposed the relatively small variations due to the changing properties of the dough itself. The latter have a

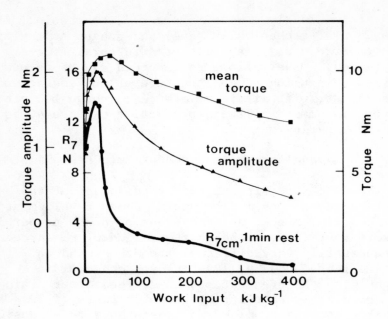

Fig. 7. Development profile constructed from dough resistance to extension after only 1 min rest (R_7), compared with mean mixer torque curve and torque trace amplitude.

much more pronounced effect on the <u>amplitude</u> of the torque trace and Fig.7 also shows the change in amplitude with work input. The peak at about 30 kJ/kg agrees with the arrival at maximum torque and then falls away sharply to give an overall change comparable in magnitude with the resistance to extension at zero rest.

Overall Assessment of Optimum Dough Development

It is clear from the above results that the rheological properties of dough are closely dependent on both the mechanical energy expended during mixing, when structural modification occurs, and the period of time allowed for relaxation afterwards, during which varying degrees of structural recovery are observed.

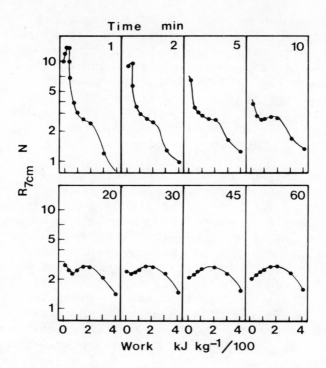

Fig. 8. Development profiles of dough, mixed in air, derived from structural relaxation curves (R_{7cm}) at rest times indicated (from 1 to 60 min.).

The overall process is summarised in Fig. 8 and it can be seen that much of the structural modification brought about by mixing lasts only for a short time. This transient structure is normally observed as a peak in torque curves produced by recording mixers and contributes to the varying forces acting on the mixer blades affecting torque trace amplitude. Such measurements suggest optimum development of doughs to occur at relatively low work levels or short mixing times. However, once dynamic mixing ceases, this transient structure is rapidly lost. Fig. 8 shows most of the mixer 'peak' structure to have disappeared by 5 min rest.

The relative importance of the slight inflection, sometimes seen in mixing curves at times well beyond 'peak', but much more apparent in structural relaxation tests, now becomes clear. It is the <u>permanent</u> structural modification of the dough which is retained during the rest period and causes this later peak - the optimum development observed in tests on rested doughs (Frazier et al., 1975).

Effect of Oxidative Improvers on Optimum Development

Using the method of compressive stress relaxation, described earlier, relaxation time values were obtained for a range of improver concentrations and work levels, analysed by computer and presented as a series of three-dimensional plots showing contour lines joining points of equal relaxation time.

Results for a typical fast-acting improver, azodicarbonamide (ADA), are shown in Fig. 9. The combined effect of oxidation and mechanical work input resulted in the position of peak dough development moving to progressively lower work levels (shown clearly by the development profile cross-sections). Good agreement was obtained with loaf volumes (Frazier et al., 1979) and this peak shift makes the relative improving effect of ADA very great at the fixed work level of 40 kJ/kg used in the Chorleywood Bread Process. However, the rapid breakdown beyond peak shows a lack of tolerance which has to be countered by blending with other ingredients such as ascorbic acid or soya lipoxygenase (Frazier et al., 1979) which modify the development profile in different ways. Further details on the action of oxidants and other improvers are contained in a very recent review (Fitchett & Frazier, 1985).

The effect of different levels of ADA on mixer torque curves during mechanical dough development is shown in Fig. 10, which may be compared directly with development profiles from the stress relaxation contour plot in Fig. 9. It is immediately evident that relatively little change occurs in the shape of the torque curves with increasing levels of ADA. However, since peak stress relaxation and optimum loaf volume move to lower work levels with increasing ADA,

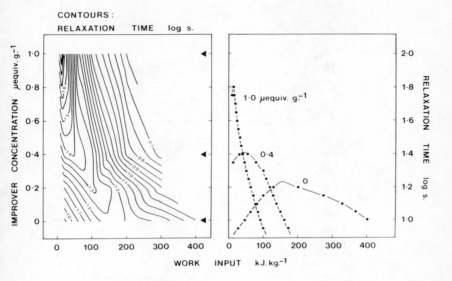

Fig. 9. Combined effect of ADA concentration and mechanical work input on the development of dough structure as determined by compressive stress relaxation measurement. Left: Three dimensional plot representing the development surface by contour lines joining points of equal relaxation time. Right: Development profile cross-sections taken at three levels of ADA addition, arrowed on contour plot (from Frazier et al., 1979).

a situation eventually arises when optimum development and peak torque coincide (in this case at about 0.6 mEquiv ADA/kg flour or approx. 35 ppm at a work input of about 20 kJ/kg). This movement of optimum development, dependent on improver concentration, almost certainly explains the many literature claims for optimum mixing to occur at peak torque or just beyond (Kilborn & Tipples, 1979).

It must be emphasised that successful use of the torque or consistency peak for the purpose of predicting optimum dough development, as in the GRL baking procedure (Kilborn & Tipples, 1979), is entirely dependent on the particular choice of improver level and type.

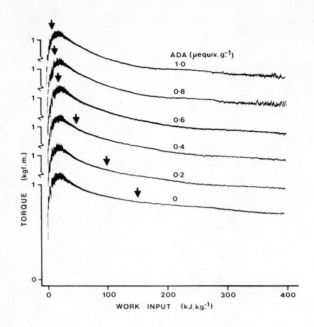

Fig. 10. Compudomixer torque curves for doughs mixed at a constant rate of work input of 20 kJ/kg.min with different levels of ADA. The position of optimum development, as determined from stress relaxation time measurements, is arrowed on each curve (from Frazier, 1979).

Similarly, where a fixed level of work input is recommended, such as 40 kJ/kg (5Wh/lb) in the Chorleywood Bread Process, attainment of optimum development at this work level is dependent on selected improvers (for the system used in Fig. 9, 0.45 mEquiv ADA/kg flour or 26 ppm would be required to achieve optimum development at 40 kJ/kg).

Rheological tests which allow for the time-dependence of wheat flour doughs and are made after a rest period equivalent to final proof, are not restricted to these special cases and provide an indication of the optimum development for all processes.

Variety Considerations: A Wheat Gluten Quality Index

Although much is now known about the interaction between improver effect and mechanical work during dough development, the mixing behaviour of wheat flours is observed to vary considerably according to the individual varieties of wheat present in the flour grist. Consequently, predictions of flour quality and improver requirement can only be determined by a study of the dough development characteristics and improver response of those pure varieties.

The method of compressive stress relaxation testing, as described earlier, can be effectively applied to a study of the development characteristics of doughs mixed from pure wheat varieties. Figure 11 shows the development profiles, in the absence of added improvers, for four UK wheats together with a strong Canadian Western Red Spring wheat. These development profiles are not merely dependent on protein quantity, but are significantly influenced by the protein 'quality' which is mainly determined genetically for the different varieties.

The development profiles classify the five flours into their order of bread baking quality, the strongest flour showing the highest relaxation time. A single numerical value, that effectively classifies each variety in terms of its development potential, can be accurately measured from the relaxation time at the peak or plateau of each curve. Multiplied by ten for convenience, the log. RT value then forms an additive quality scale ranging from 1 to 20, with most wheats falling in the range 9 to 18. Values of this Gluten Quality Index for the flours in Figure 11 are thus:- Mardler 9.5; Kador 11; Flanders 12; Avalon 14; CWRS 17-18.

Whilst the mixing characteristics of wheat flours may be usefully studied in the absence of added improver, to establish their inherent gluten quality, it is very important when fully evaluating the commercial usefulness of particular varieties to consider the response of the flour to the improvers that will be used in the bakery. This response

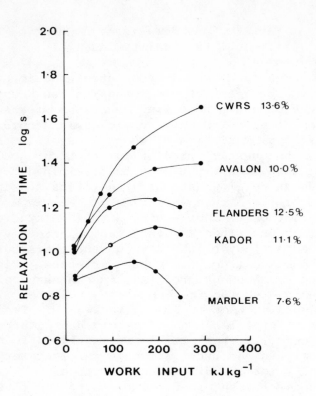

Fig. 11. Dough development profiles, without improvers, for different wheat varieties at indicated protein contents (%) (from Fitchett & Frazier, 1985).

determines how much of the gluten quality potential (usually revealed at high work levels) can be effectively realised at CBP work input or lower by improver interaction changing the shape of the development profile. By determining development profiles (stress relaxation time vs work input) for different varieties and improver components or mixtures, it is possible to predict the baking behaviour of most flour-improver combinations and so gain information very pertinent to their optimum commercial use.

SUMMARY

Torque-recording mixers, although widely used for monitoring comparative dough consistency and development, provide information which is strongly dependent on short-term structural modification. During a subsequent rest period, much of this transient structure is lost through structural relaxation or recovery. Only after 30 min rest or more does the permanent structural modification become clearly apparent, usually showing an optimum development condition at a much higher work level or longer mixing time than that suggested by the initial, short-term torque peak. This time-dependent transformation between the two development peaks has been clearly followed in rheological tests.

In the presence of oxidising improvers, the permanent structural development peak occurs progressively earlier in the mixing cycle, at lower work input, depending on improver type and level, and under appropriate conditions the transient and permanent peaks may coincide. Only in these special cases will optimum development for baking be predicted from mixer torque peaks. All other conditions require tests on rested doughs to provide an accurate prediction of optimum development.

Such tests, using compressive stress relaxation, readily enable single-variety wheat flour samples to be examined for potential baking strength and a simple 'gluten quality index' is proposed.

ACKNOWLEDGEMENTS

The authors thank Mrs F.A. Brimblecombe and Mrs M. Tuck for structural relaxation work and statistical analysis, S.M. Brown and P.A. Bailey for technical assistance.

LITERATURE CITED

Bushuk, W., and Mehrotra, V.K. 1977a. Studies of water binding by differential thermal analysis. I. Dough studies using the boiling mode. Cereal Chem. 54:311.

Bushuk, W., and Mehrotra, V.K. 1977b. Studies of water binding by differential thermal analysis. II. Dough studies using the melting mode. Cereal Chem. 54:320.

Daniels, N.W.R., 1975. Some effects of water in wheat flour doughs. Page 573 in: Water Relations of Foods. R.B. Duckworth, ed. Academic Press, London.

Daniels, N.W.R., and Frazier, P.J. 1978. Wheat proteins - physical properties and baking function. Page 299 in: Plant Proteins. G. Norton, ed. Butterworths, London.

Davies, R.J., and Webb, T. 1969. Calorimetric determination of freezable water in dough. Chem. Ind. (London) 1138.

Dempster, C.J., Hlynka, I., and Winkler, C.A. 1952. Quantitative extensograph studies of relaxation of internal stresses in non-fermenting bromated and unbromated doughs. Cereal Chem. 29:39.

Dempster, C.J., Hlynka, I., and Anderson, J.A. 1953. Extensograph studies of structural relaxation in bromated and unbromated doughs mixed in nitrogen. Cereal Chem. 30:492.

Dempster, C.J., Hlynka, I., and Anderson, J.A. 1955. Influence of temperature on structural relaxation in bromated and unbromated doughs mixed in nitrogen. Cereal Chem. 32:241.

Fitchett, C.S., and Frazier, P.J. 1985. Action of oxidants and other improvers. In: Chemistry and Physics of Baking - Materials, Processes, Products. J.M.V. Blanshard, P.J. Frazier, T. Galliard, eds. Royal Society of Chemistry, London. (Symposium proceedings, in press).

Frazier, P.J., Leigh-Dugmore, F.A., Daniels, N.W.R., Russell Eggitt, P.W., and Coppock, J.B.M. 1973. The effect of lipoxygenase action on the mechanical development of wheat flour doughs. J. Sci. Food Agric. 24:421.

Frazier, P.J., Daniels, N.W.R., and Russell Eggitt, P.W. 1975. Rheology and the continuous breadmaking process. Cereal Chem. 52:106r.

Frazier, P.J., Brimblecombe, F.A., Daniels, N.W.R., and Russell Eggitt, P.W. 1979. Better bread from weaker wheats - rheological considerations. Gehreide, Mehl u Brot 33:268.

Frazier, P.J. 1979. A basis for optimum dough development. Baking Ind. J. 12(1): 20.

Halliwell, W. 1898. Flour and flour tests. Page 179 in: The Technics of Flour Milling: a Handbook for Students. The Northern Publishing Co. Ltd., Liverpool.

Heaps, P.W., Webb, T., Russell Eggitt, P.W., and Coppock, J.B.M. 1968. The rheological testing of wheat glutens and doughs. Chem. Ind. (London) 1095.

Kilborn, R.H., and Tipples, K.H. 1979. The effect of oxidation and intermediate proof on work requirements for optimum short-process bread. Cereal Chem. 56:407.

Launay, B., and Bure, J. 1974a. Etude de certaines proprietes rheologiques des pates de farine; influence de la duree du petrissage sur ces proprietes. Dechema Monographien 77:137.

Launay, B., and Bure, J. 1974b. Stress relaxation in wheat flour doughs following a finite period of shearing. Cereal Chem. 51:151.

Muller, H.G., Williams, M.V., Russell Eggitt, P.W. and Coppock, J.B.M. 1961. Fundamental studies on dough with the Brabender Extensograph. I - Determination of stress-strain curves. J. Sci. Food Agric. 12:513.

Munz, E. and Brabender, C.W. 1940a. Prediction of baking value from measurements of plasticity and extensibility of dough. I. Influence of mixing and moulding treatments upon physical dough properties of typical American wheat varieties. Cereal Chem. 17:78.

Munz, E. and Brabender, C.W. 1940b. Extensograms as a basis of predicting baking quality and reaction to oxidising agents. Cereal Chem. 17:313.

Webb, T., Heaps, P.W., Russell Eggitt, P.W. and Coppock, J.B.M. 1970. A rheological investigation of the role of water in wheat flour doughs. J. Food Technol. 5:65.

RHEOLOGY OF FERMENTING DOUGH

R. Carl Hoseney

Department of Grain Science and Industry
Kansas State University
Manhattan, KS 66506

Contribution No. 86-87-A
Kansas Agricultural Experiment Station

Bread dough is a viscoelastic material. It is essentially nonlinear and appears not to have a yield point (Hibberd and Parker, 1975). From a compositional standpoint bread dough is very complex. In addition, it is made up almost entirely of biological material that would be expected to vary from one source to another and from one time to another.
In an attempt to simplify the system most rheologists have only studied flour-water systems rather than complete dough. While this has greatly simplified the system it has not, by any means, produced a simple system to study. Flour in itself is a complex biological material that can vary quite widely. Two items beyond the variation within the flour should be discussed. First, the level of water added to flour can be quite important, together with the fact that different flours may require different amounts of water (Hoseney and Finney, 1974). This strikes most rheologists as a strange approach. Equally strange to most rheologists is the concept of mixing to an optimum

mixing time rather than for a constant time (Fig. 1). However, it is imparative that dough be prepared with optimum water and be mixed to an optimum point. A dough that is significantly under or overmixed is rheologically quite different from an optimum dough.

A flour-water dough is a multiphasic system with a continuous hydrated protein phase and two discontinuous phases, starch and air. It is relatively easy to study because it is relatively stable. The changes that are occurring in the dough are occurring at a relatively slow rate. This is in contrast to the more rapid changes found with fermenting doughs. The presence of yeast in the dough gives another discontinuous phase, however, the relatively small amount of yeast probably is of little significance. The products of yeast fermentation are undoubtedly of much greater significance.

As fermentation proceeds the level of oxygen in the dough drops rapidly and the fermentation becomes anaerobic. The carbon dioxide produced has two major effects. First the pH drops rapidly as the carbon dioxide dissolves in water forming carbonic acid. The rheology of dough is affected by pH. Another important aspect of the production of carbon dioxide is that the aqueous phase soon becomes saturated and then the carbon dioxide finds its way into the preexisting air cells. As a result the dough is leavened (Hoseney, 1984). The leavening process changes the geometry of the dough piece. In addition, it is possible that the large amount of gas retained in the dough changes dough rheology based on the increased surface produced. This will be discussed in more detail later. Those changes make the study of dough rheology in a fermenting dough a challenging problem.

Besides the above described problems, we should not ignore the other changes that occur in dough as a result of oxidants. The action of these reagents can be shown in nonfermenting dough. Also, the possibility that yeast may be responsible for changes in dough rheology cannot be ignored.

Reports in the literature of attempts to meas-

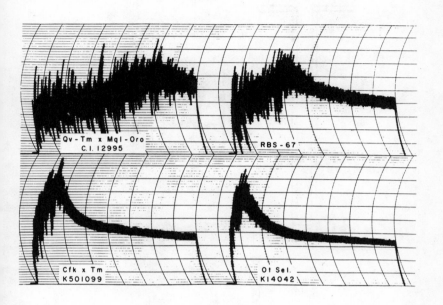

Figure 1. Mixogram of flours varying in mixing time

ure the rheology of fermenting dough are not numerous. Perhaps the first successful attempt was by Bailey (1955). Using the device shown in Figure 2 to deflate a fermented dough at the end of proofing, he showed that the pressure in the gas cells was an average of 1.032 atmospheres. In considering a fermenting dough piece, it is obvious that the pressure inside the gas cell must be higher than atmospheric pressure or the dough would not rise.

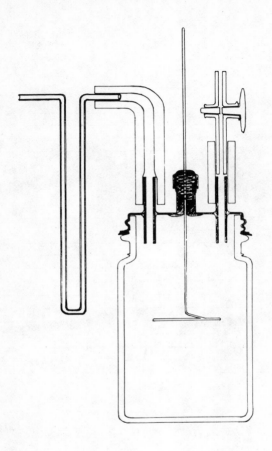

Figure 2. Device used by Bailey (1955) to measure the pressure in fermenting dough.

It is also obvious that the system must approach an equilibrium with the excess pressure equalling the resistance of the cell walls to expansion. The fact that the excess pressure is small shows that the resistance to expansion is also small when measured over a long time span. Stated in other words, when the dough is allowed to relax the viscous flow is large and the elastic property is small.

Matsumoto and his coworkers (Matsumoto et al 1971, Matsumoto et al 1973, Matsumoto et al 1974, Matsumoto and Mita 1983, and Mita and Matsumoto 1978) has studied the rheology of fermenting dough by the bottom pressure of fermenting dough in a cylinder and from the stress exerted to a horizontal wire ring which penetrates the surface membrane of the same fermenting dough. The basic apparatus used is shown in Figure 3. A later model was automated (Matsumoto et al 1973). The apparatus was shown to be effective at measuring the pressure above the rubber membrane (A). When dough (D) was placed above the membrane a pressure was developed. As pointed out by the authors, a number of factors affect the pressure reading. The size of the cylinder, and therefore, presumably the friction of the dough to the cylinder wall. The strength of the dough with stronger dough giving a higher pressure. It was also shown that the amount of dough in the cylinder affected pressure. Rheologically active agents such as the oxidant $KBrO_3$, increased the pressure while reducing agents such as cysteine reduce the pressure (Matsumoto et al 1975). Thus, it appears clear that the technique is measuring the rheological properties of fermenting dough. The use of the term internal pressure may be misleading as it is not clear that the pressure inside the dough is actually being measured (Bloksma 1981).

In a later refinement of the technique (Matsumoto and Mita 1983) a wire ring was positioned above the dough and as the dough expands, the stress on the ring is measured. When the stress on the ring was plotted versus the bottom pressure a straight line with a slope of 1 was obtained (Fig. 4). The author interpreted that data to support the concept that fermenting dough acts as a multi-gas cell structure and that the stress at the top of the dough goes through the gas cells and reaches the bottom without loss. Clearly the stress on the ring is a measure of the rheology of the fermenting dough; $KBrO_3$ increases the stress and cysteine decreases the stress. Given the stress on the ring the authors calculated the tension of the dough

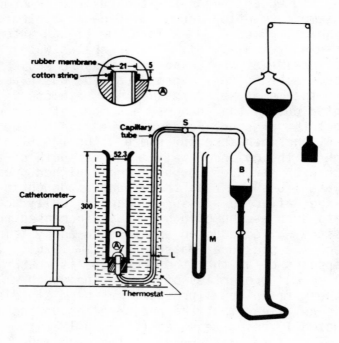

Figure 3. Device to measure the pressure of yeasted dough (Matsumoto et al. 1971)

surface (Table 1).

Carlson and Bohlin (1978) have presented calculations suggesting that the surface free energy is an important contributor to elasticity of wheat flour dough. They estimated that as much as 80% of the elastic energy is attributable to this source. Blosksma (1981) has criticized their assumptions and has presented an alternate model. That model pre-

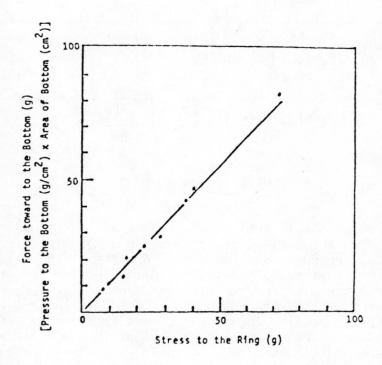

Figure 4. Response of stress on the ring at the surface to the pressure of the dough on the bottom (Matsumoto and Mita 1984)

dicts that the contribution of surface free energy is of little importance.

A simple test to measure the rheological properties of fermenting dough involved measuring the width and height of a moulded dough cylinder that had been rested in a high humidity chamber (Hoseney et al 1979). The logic of the test, essentially a

Table 1. Tension of the dough membrane on a ring at the surface. Adapted from Matsumoto and Mita 1984.

Dough	Maximum Stress g	Tension dynes/cm
Control	69.9	5.41
$KBrO_3$ (30 ppm)	86.1	6.67
Cysteine (30 ppm)	53.8	4.17

$$T(dynes/cm) = \frac{Stress \times 980}{2\pi r + 2\pi R}$$

r = inner radius
R = outer radius

creep test, is outlined in Figure 5. Three major forces are acting on a fermenting dough piece; gravity, pressure, and cohesion. Gravity is the major force that makes the dough flow. Pressure resulting from the gas produced within the dough is exerted at every part of the dough and as a result the dough expands equally in all directions to increase its volume. Thus, the ratio of height to width is independent of the increase in volume. The cohesive forces within the dough resist expansion and limits flow. The cohesive forces are actually made up of elastic and viscous components.

One of the advantages of this test is that the force on the dough and the rate of deformation are the same as those during baking. When a flour-water system is given a "fermentation" time, the viscous flow property increases (Fig. 6), but when yeast is added to the flour-water system the viscous flow becomes less. The effects of the bread dough ingredients on the spread ratio of a dough were studied by eliminating the ingredients one at a time from the complete formula. The spread ratio was measured after 0 and 180 min of fermentation (Table

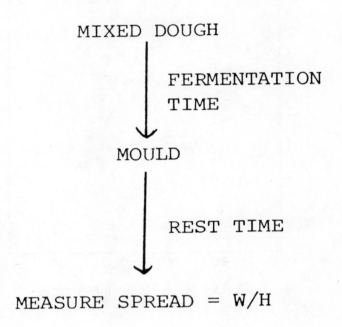

Figure 5. Scheme used to measure the rheological properties of fermenting dough (Hoseney et al. 1979)

2). At zero fermentation, sugar was the only ingredient that had a major effect on the spread ratio. The spread ratio was reduced when sugar was not present in the dough. When sugar dissolves in water it displaces 0.6 cc/grams. Thus, 6 g of sugar would equal 3.6 cc of solution. This would be enough to effect the flow properties of the dough.

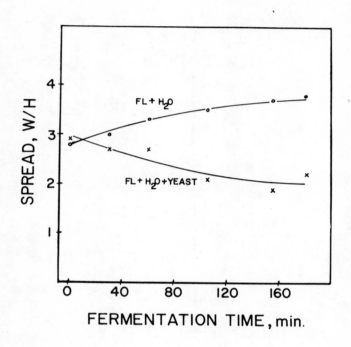

Figure 6. Effect of yeast on the spread ratio of a fermenting dough (Hoseney et al. 1979)

After 180 min of fermenation only yeast and $KBrO_3$ were shown to effect the spread ratio. The oxidant would be expected to have an effect on the rheology while yeast was shown to be effective in Figure 6.

The spread ratio value of 1.5 after 180 min of fermentation is essentially equal to no flow. A cylinder of dough would have a ratio of 1 but when placed on a surface it elastically deforms and gives

Table 2. Effects of baking ingredients on the spread ratio of dough.

Ingredients	Spread - W/H, Fermentation	
	0 min	180 min
Control	3.0	1.5
Sugar	2.4	1.5
Salt	2.8	1.6
Nonfat dry milk	3.1	1.5
$KBrO_3$	2.9	1.7
Yeast	3.0	2.5
Shortening	2.8	1.5
Malt	2.8	1.5

a ratio of about 1.5. Therefore, under the force of gravity a fully fermented dough is elastic and does not show viscous flow properties. The viscous properties may be seen if we increase the strain on the dough.

It appears to be particularily important to note that sugar had no effect on the spread ratio after 180 min of fermentation. This shows that the amount of gas in the dough and, thus, its volume does not affect the spread ratio. This data would support the conclusion of Bloksma (1981) discussed above.

Oxidizing agents other than $KBrO_3$ all decreased the spread ratio as would be expected and reducing agents increased the spread ratio also as expected. Yeast was the major ingredient contribution to rheological changes in a fermenting dough. It was not clear how yeast changed dough rheology. To check if the products of fermentation were responsible for the change, a liquid ferment system was used. After fermentation the mixture was centrifuged to separate the yeast cells from the products of fermentation. The effect of those products on dough rheology could be determined by adding them to flour and mixing into a dough. The spread test showed that the products of fermentation did not change dough rheology. Therefore, the rheological changes in fermenting dough are apparently caused by the yeast itself. Whether this is an enzyme of the yeast or an intermediate product produced during fermentation is not clear.

The spread test also has disadvantages. Two major ones are the time required to run the test and the fact that fundamental rheological data are not obtained.

The rheological changes that occur when a dough is heated have not been widely studied. The major work reported appears to have been from the Lord Rank Research Center of RHM Research Limited in High Wycombe, U.K. In a series of articles (Schofield et al 1983a, LeGrys et al 1981, and Schofield et al 1983b) the effects of heating on wheat gluten were reported. They showed that the baking performance of wheat gluten progressively declined as a result of heating (Table 3). Most

Table 3. Effect of heating wet gluten and the addition of cysteine on its baking performance. Data from Schofield et al 1983a.

Temperature	Loaf Volume	Cysteine Added	Loaf Volume
$°C$	cc	ppm[a]	cc
unheated	200	0	110
60	147	207	151
70	115	829	173

[a] based on gluten weight, added to gluten heated to $70°C$.

of the functionality was a decrease in extractability of the gluten proteins. Chromatographic data showed that most of the change in extractability was because of changes in the glutenin proteins. Heating to temperatures higher than 75°C decreased the solubility of gliadin proteins.

The heating appeared to trigger a sulphydryl/disulfide interchange that apparently was responsible for the decrease in solubility and the poorer baking results. Surprisingly, most of the loss in baking performance could be recovered by adding cysteine to the gluten (Table 3). This infers that the heating had no adverse effects upon gluten other than the polymerization resulting for sulphydryl/disulfide interchange. This result is consistent with the lack of a denaturation peak when gluten was examined by differential scanning calorimetry (Eliasson and Hegg 1980, and Schofield et al 1983a, Arntfield and Murray 1981).

LITERATURE CITED

Arntfield, S. D. and Murray, E. D. 1981. The influence of processing parameters on food protein functionality. I. Differential scanning calorimetry as an indicator of protein denaturation. J. Inst. Can. Sci. Tech. 14:289.

Bailey, C. H. 1955. Gas pressure in fermented doughs. Cereal Chem. 32:152.

Bloksma, A. H. 1981. Effect of surface tension in the gas-dough interface on the rheological behavior of dough. Cereal Chem. 58:481.

Carlson, T. and Bohlin, L. 1978. Free surface energy in the elasticity of wheat flour dough. Cereal Chem. 55:539.

Eliasson, A. -C. and Hegg P. -O. 1980. Thermal stability of wheat gluten. Cereal Chem. 57:436.

Hibberd, G. E. and Parker, N. S. 1975. Measurement of fundamental rheological properties of wheat flour doughs. Cereal Chem. 52:1r.

Hoseney, R. C. 1984. Gas retention in bread

doughs. Cereal Foods World. 29:305.

Hoseney, R. C. and Finney, P. L. 1974. Mixing - a contrary view. Baker's Digest. 48(1):22.

Hoseney, R. C., Hsu, K. H. and Junge, R. C. 1979. A simple spread test to measure the rheological properties of fermenting dough. Cereal Chem. 56:141.

LeGrys, G. A., Booth, M. R. and Al-Baghdadi, S. M. 1981. The physical properties of wheat proteins. pp. 243. In: Cereals: A renewable resource-theory and practice. Y. Pomeranz and L. Munck, eds. AACC. St. Paul.

Matsumoto, H., Nishiyama, J. and Hlynka, I. 1971. Internal pressure in yeasted dough. Cereal Chem. 48:669.

Matsumoto, H. Nishiyama, J. and Hlynka, I. 1973. Internal pressure in yeasted dough II. Cereal Chem. 50:363.

Matsumoto, H. and Mita, T. 1983. Penetrating stress of a wire ring to yeasted dough. Cereal Foods World. 28:563. Abstract #85.

Matsumoto, H. and Mita, T. 1984. Rheology of yeasted dough. Sym. on Advances in Baking Science and Technology. Department of Grain Science and Industry, Manhattan.

Matsumoto, H., Nishiyama, J., Mita, T. and Kuninori, T. 1975. Rheology of fermenting dough. Cereal Chem. 52:82r.

Matsumoto, H., Ono, H. and Mita, T. 1974. Relaxation of pressure in dough. Cereal Chem. 51:758.

Mita, T. and Matsumoto, H. 1978. Relationship between internal pressure and bubble size in fermenting dough. J. Agric. Chem. Soc. Japan (Japanese). 52:111. Original not seen.

Schofield, J. D., Bottomley, R. C., LeGrys, G. A., Timms, M. F. and Booth, M. R. 1983a. Effects of heat on wheat gluten. p. 81. Proceedings of the 2nd International workshop on gluten proteins. Eds. A. Graveland and J. M. E. Moonen.

Schofield, J. D., Bottomley, R. C., Timms, M. F. and Booth, M. R. 1983b. The effect of heat on wheat gluten and the involvement of sulphydryl-disulfide interchange reactions. J. Cereal Sci. 1:241.

RHEOLOGY OF SOFT WHEAT PRODUCTS

J. Loh
Central Research Division
General Foods Corporation
Tarrytown, New York 10591, USA

INTRODUCTION

The importance of texture in consumer acceptance of foods has been well established (Szczesniak and Kleyn, 1963; Szczesniak, 1971 and 1972; Szczesniak and Kahn, 1971; Yoshikawa et al. 1970a and b). It is also known that the relative importance of texture may vary depending upon the type of food (Bourne, 1982). Baked goods made from soft wheat flour, such as cakes and crackers, belong to the category of foods in which the textural quality is considered to be "important", if not "critical", to the overall acceptability.

There are many diverse reasons for the importance of rheological properties (Szczesniak, 1969; Sherman, 1970; Van der Tempel, 1980). However, the successful application of rheology to any product depends primarily on the ability to accurately measure the properties meaningful to the intended purpose. Objective quantification of specific rheological parameters remains a challenge to food rheologists.

Soft wheat flour, characteristically low in gluten, water absorption capacity and size of granulation, is used to make cakes, cake doughnuts, cookies, crackers, wafers, pretzels and similar products. They represent a gamut of texture ranging from soft/springy (e.g. in cakes) to brittle/ crunchy (e.g. in crackers). The ability to measure these textural characteristics instrumentally for objectivity, time/labor effectiveness and good reproducibility has become a necessity. Due to the large number of products and the current trend to employ universal testing instruments, the options in testing methodology are numerous and the selection of proper methodology for specific product/purpose becomes critical.

This paper will define the important textural parameters of soft wheat products, review the applicable published test methods, provide general guidelines for the selection of the test mode and test condition, and discuss recent trends in rheological characterization of baked soft wheat products.

TEXTURAL PARAMETERS

According to the classification given by Szczesniak (1963a), textural chracteristics can be grouped into mechanical, geometrical and those related to fat and moisture content. Rheology is manifested in the mechanical response of the food to applied stress or strain. Organoleptically, these characteristics are perceived by the pressures exerted on fingers, teeth, tongue and other interior surfaces of the mouth. The stress/ strain can be applied both non-destructively and destructively by bending, compressing, tearing, biting, masticating and/or swallowing. Texture is detected in the "as is" product, in the wetted/ masticated mass ready for swallowing and in any state in between. The mouthfeel properties of saliva wetted products (grossly ignored by researchers) involve all three basic groups of textural characteristics. They may be described as viscousness, sliminess, smoothness, mouthcoating, ease of swallowing, etc.

Table 1 lists the popular terms of mechanical nature used to describe the texture of soft wheat products. Using vanilla cookies as an example, textural characteristics other than mechanical ones including smoothness, density, uniformity and dryness were mentioned by Civille and Liska (1975). Figure 1 shows the typical stress/strain relationship and indicates the proper grouping of the parameters based on the degree of deformation. The terms "firmness" and "strength" denote different properties and should not be used interchangeably. On the other hand, "compressibility" (= 1/firmness) and "firmness", used in literature as separate characteristics, relate to the same property.

RHEOLOGICAL TESTS APPLICABLE TO SOFT WHEAT PRODUCTS

Instrumental rheological tests for foods can be generally divided into three groups: Fundamental, empirical and imitative (Szczesniak, 1963a). The literature on soft wheat products contains mostly empirical methods. The use of fundamental and imitative tests for characterization of baked goods is relatively limited. Recently acoustic (non-rheological) methods have been employed for crispness quantification and are applicable to brittle baked goods.

(A) Fundamental Tests

Fundamental tests measure well-defined rheological parameters such as viscosity and elastic modulus. These parameters (expressed in basic units) reflect the structural properties of the material on molecular and microscopic levels. Fundamental tests are more popular in studies on batter and dough to describe time-dependent rheological behavior. Most foods possess both

TABLE 1
Important Textural Parameters of Soft Wheat Products

(A) In "As Is" Form

Parameter	Related Terms
Hardness	Softness, firmness, compressibility, bendability.
Springiness	Elasticity, recovery from deformation.
Fracturability	Strength, brittleness, crispness, crunchiness, snappability.
Chewiness	Tenderness, toughness.
Adhesiveness	Stickiness, tackiness.
Cohesiveness	Crumbliness, type/rate of breakdown.

(B) In "Masticated/Wetted Form

Parameter	Related Terms
Viscosity	Viscous, thick, thin, ease of swallowing.
Gumminess	Shortness, pastiness, breakdown.
Adhesiveness	Mouthcoating, molar packing, stickiness.

solid-like elastic and fluid-like viscous properties and are described as "viscoelastic". Three common tests ued to characterize viscoelasticity are creep (e.g. Shama and Sherman, 1968), relaxation (e.g. Vovan et al, 1982) and dynamic strain application.

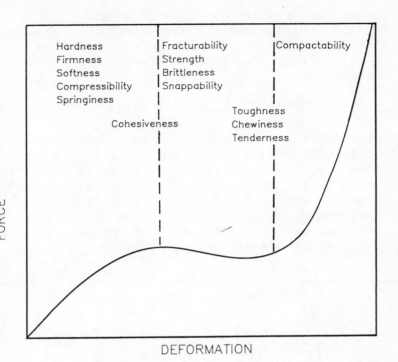

Fig. 1. Relationship between degree of deformation and detection of textural characteristics.

Expensive instruments and trained operators are usually needed, and data interpretation is often difficult due to lack of knowledge on the precise relationship between rheology, structure and functional properties of a complex product such as food. Szczesniak (1963b) indicated that "....fundamental tests serve the greatest value to the food technologist by providing bases for the development of more meaningful empirical tests." Information obtained with fundamental tests may be critical in avoiding costly mistakes resulting from misleading empirical tests/test conditions. This is illustrated in Fig. 2, which demonstrates that the viscoelastic response of pound cake is highly strain dependent, and that the values of the elastic (G') and viscous (G") modulus decrease with increasing deformation above 1% strain. The nonlinear viscoelastic behavior of pound cake means that different results will be obtained from empirical tests involving different degrees of deformation.

(B) Empirical Tests

Empirical tests measure parameters which are often poorly defined, but which are found in practical experience to be correlated to textural quality of the product. In these methods, the mode of force or strain application can be simple (e.g. compression, tension, penetration, bending, shearing, cutting, extrusion) or complex (i.e. encompassing several different modes).

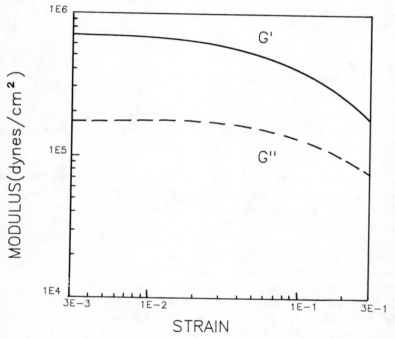

Fig. 2 Strain dependence of storage modulus (G') and loss modulus (G") of cake (sample was tested at 1 Hz and 25°C using 25 mm diameter parallel plates).

Firmness: Firmness is traditionally measured to assess the textural quality of breadcrumbs; these tests are, in general, applicable to soft/spongy cakes. The measurement involves applying a constant load and measuring the extent of deformation in a given time or applying a constant deformation and measuring the product's resistance.

Platt (1930) used a simple compressimeter consisting of a balance with one of its pans attached to a cylindrical plunger. Compression of crumb during 1 min. and the degree of recovery after load removal were determined. This test was later improved by Platt and Powers (1940) by studying the effects of slice thickness, stress level and rate of deformation.

Measurement using the Baker compressimeter (AACC, 1947) is a Standard Cereal Laboratory Method. Pence and Standridge (1958) equipped it with a pressure plate 1-1/4 in. square and used it on 1 in. thick sections of cake crumb. Force readings were taken at 4 mm compression. Bechtel and Kulp (1960) used this instrument to study the effect of freezing and thawing on the firmness of doughnuts.

Compression testing was also used for measuring hardness of dry/crunchy products e.g. pretzels (Bourne and Comstock, 1981) and cookies (Hutchinson et al, 1977).

Panimeter, another balance-type compressimeter, was described and used by Robson (1966) to measure crumb firmness by compressing a sample under conditions of constant stress. Robson pointed out that compressibility must be measured under non-destructive conditions. In view of the non-linear behavior of cake structure, he felt that results would be more reliable under the condition of constant strain (i.e. firmness measurement).

Compressibility or firmness has also been quantified using more precise instruments, such as the Kramer Shear Press equipped with a succulometer cell (Brown and Zabik, 1967) and Instron equipped with a parallel plates fixture (Gay and Wren, 1968; Vovan et al 1982). Babb (1965) developed an instrument for the rapid evaluation of the compressibility of bakery goods of uniform thickness.

Using the same instrument, breaking force of biscuits of different moisture content has been measured by substituting the compression block with a rounded plunger.

Springiness: Springiness is an important textural parameter for cakes and is quantified as the degree of sample recovery upon load removal. Robson (1966), using the Panimeter, measured springiness of crumb as % recovery during decompression. The value of springiness measured is highly dependent on the degree of compression and should be measured under small deformation conditions (Fig. 1).

Strength: The mechanical strength of any given material is affected by the method of stress application. Normally, "strength" refers to rupture strain in tension. It is also called in the literature "toughness/tenderness", "snapping force", "hardness", etc. Shortometer (Bailey, 1930 and 1934) based on the method established by Davis (1921) traditionally has been used for the evaluation of tenderness of pastry/cakes and shortness or snapping force of crackers and cookies. The sample is laid over two parallel supporting rods 6.5 mm in diameter and 40 mm apart, and force is applied through a flat metal rod attached to a motor. Failure or fracture strength is measured (Fisher, 1933; Harvey, 1937; Lowe et al. 1938; Briant et al. 1957; Hirahara et al. 1961; Matthews and Davison, 1963; and Stinson and Husk, 1969). Quality of correlation with sensory tenderness measurement was questioned by several reseachers (e.g. Harvey, 1937 and Horstein et al. 1943). Breaking strength of crunchy low-moisture baked goods has also been determined by bending/snapping using the Instron Universal testing machine equipped with a triple beam test cell (Rammell et al. 1971; Robinson, 1971; Katz and Labuza, 1981).

Penetrometer is another simple instrument used commonly to assess the strength of baked goods (Smejkalova, Z., 1974; Babb, 1965, Funk et al. 1969; Morandini et al. 1972; Choishner et al. 1983). A probe or indentor is generally used and the depth of penetration at a definite time after loading a constant weight is recorded. The greater the penetration, the more tender the product. Conversely, force at a constant penetration depth may also be used to indicate tenderness. The smaller the force, the more tender the product. Theory and application of puncture testing was given by Bourne (1979b).

Toughness or strength of baked goods has also been measured by tensile testing. Platt and Kratz (1933) used a device to extend a piece of 4x6x1 in. cake crumb at a rate of 200 g/min and followed the development of toughness in cakes during staling. Brown and Zabik (1967) used the Kramer Shear Press and a special clamp attachment to measure the tensile strength of cakes. Due to the difficulties in sizing, shaping and clamping the test specimen, the tensile test is a less-preferred test for baked goods.

Another popular way of measuring toughness is by shearing. This type of test (e.g. shear press) rarely applies simple shear to the sample. In fact, a complex mode of force application is used including one or more of the following modes: shear, compression, penetration, cutting, tension, tearing, even extrusion. Many researchers have used Kramer Shear Press equipped with a standard 10 blade shear/compression cell to measure the tenderness/toughness of cakes or cookies [Szczesniak et al (1970), Brown and Zabik (1967), Funk et al (1969), Stinson and Huck (1967), Gruber and Zabik (1966) and Matthews et al (1963)]. Zabik et al (1979) used the shear press with a single blade cell to determine breaking strength of sugar-snap cookies. Schaller and Mohr (1976) described a shear device and used it to characterize the shear strength of waffles based on recorded maximum force and consideration of shearing area.

The ease in cutting or sawing of the sample was used to indicate the texture of baked goods. Robson (1966) described a Rotary Cutter to assess the crumb strength from the force developed when a rotating blade cut a spiral path in a cake sample. Method and device used to determine the hardness of biscuits by measuring the time required to cut into a stack of biscuits with a circular saw blade was reported by Wade (1968).

Adhesiveness: Adhesiveness or stickiness is important in cakes. The Struct-O-Graph (C.W. Brabender, South Hackensack, N.J.) was used by Gaines (1962) to measure stickiness of cake crumb. Cake made from chlorinated flour was found to be more sticky than the control cake (Gaines and Donelson, 1962).

Crispness: Crispness, brittleness or fracturability in friable products can be measured directly by mechanical means. Force or work to rupture under bending or compression and the suddeness of the rupture are generally good indications of crispness. Szczesniak (1963b) equated crispness with low cohesiveness and a relatively high hardness. The sound accompanying the mechanical failure is known to contribute to the overall crispness perception. This was first studied by Drake (1963, 1965) and reviewed by

Vicker (1976). Mohamed et al (1982) stressed the important role of food sound in sensory crispness. Jowitt and Mohamed (1980), using a constant loading rate device and a sound recording system, characterized the fracture behavior of sponge fingers, wafer biscuits, crisp toasts and ice cream wafers.

<u>Mouthfeel Properties</u>: All the instrumental tests described in the literature relate to the characterication of the "as is" baked goods. In the mouth, the products are not only disintegrated mechanically on mastication, but are also mixed with saliva which softens and hydrates the structure. The viscometic property of the saliva-food mixture has an important bearing on the feel in the mouth and on swallowing characteristics.

In our own work, we have been characterizing the flow properties of cake/water slurries using a Haake rotational viscometer. Typical flow curves are shown in Fig. 3B. The flow is plastic, i.e. the slurry exhibits a yield point followed by decrease in viscosity with increase in shear rate. It should also be noted from Fig. 3 that as compared to Cake B, Cake A, which was more resistant in the shear/compression cell of the Kramer Shear Press when tested "as is", was less resistant to flow (lower in viscosity) when mixed with water. This indicates that, although initially more resistant to mastication, Cake A will soften more in the mouth and should be easier to swallow.

(c) Imitative Tests

Imitative tests applicable to baked goods are mainly of the type simulating mastication. Typical here is the General Foods texturometer composed of mechanical jaws, a strain gauge and a recording system (Friedman et al, 1962). The instrument is being made by the Zenken Co. of Tokyo and is used

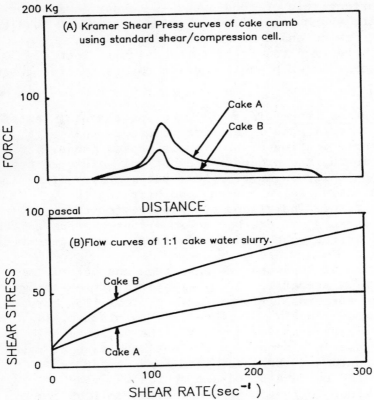

Fig. 3 Comparison of two cake samples in crumb and slurry form using shear press and rotational viscometer, respectively.

extensively in the Orient for textural characterization of foods (Tanaka, 1975). The test uses a "two bite" compression of a bite-size piece of food and a force-time curve interpretation, as illustrated in Fig. 4. During testing, the sample is placed on a flat stage and compressed twice usually in a destructive fashion by a mechanized jaw simulating the action of mastication.

Plungers of different size, shape and material of construction, degree of compression and speed of plunger motion can be used to give optimum results with different products. These quantified

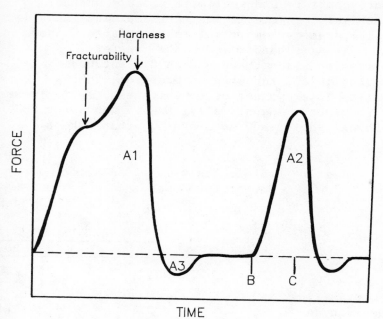

Fig. 4. A typical texture profile analysis curve of cake obtained with the General Foods texturometer; Cohesiveness= A1/A2; Springiness =BC; Adhesiveness= A3; Gumminess= Hardness x Cohesiveness; Chewiness= Gumminess x Springiness (slightly modified from Freidman et al., 1963)

parameters include hardness, cohesiveness, fracturability, springiness and adhesiveness. Excellent correlations with sensory ratings have been reported (Szczesniak et al 1963, Brandt et al 1962; Tanaka et al 1975). The same test may be performed using a universal testing machine such as an Instron (Bourne, 1968). The method is known as instrumental texture profiling analysis (TPA). Applications to baked goods have been reported by Tanaka et al (1973 a and b).

In keeping with the multi-dimensional nature of food texture, texture profile analysis provides a more complete and realistic picture of the mechanical characteristics of the product than do single-parameter empirical tests. Table 2 shows that Cake A is harder and has a greater compressive strength than Cake B. But under destructive and repeated compression, Cake A exhibits lower values of cohesiveness, springiness and chewiness. Sensorially, Cake A was described as harder, drier and more crumbly than Cake B.

Table 2
Texture Profile Analysis of Selected Butter Pound Cakes

Textural Parameters	Cake	
	A	B
Hardness (kg)	3.35	2.85
Fracturability (kg)	0.50	0.23
Cohesiveness	0.49	0.61
Springiness (cm)	1.65	2.95
Gumminess (kg)	1.64	1.74
Chewiness (kg.cm)	2.71	5.13

PRINCIPLES IN TEST SELECTION

The general principles involved in the selection of a rheological test methodology have been discussed by many researchers (e.g. Szczesniak 1963, 1966, 1973; Kramer and Twigg 1959; Szczesniak and Torgenson 1965 and Heiss and Witzel 1969). To select a proper rheological test for specific products, one must first consider the purpose of the test (e.g. consumer acceptability, engineering design, or structural analysis). The selected test should not only serve the intended purpose, but the chosen test mode and test conditions should reflect as closely as practically possible the actual stress/strain conditions to which the product is subjected in practice. When comparing competitive products and defining targets for product development, a more descriptive TPA may be more suitable than a single parameter empirical test. In selecting the test mode, the decision is based primarily on two factors: the specific rheological parameters to be measured and the nature of the products. The rule of thumb is to look for relevance if a close simulation is not feasible. For example, if the tearing strength of a cake is of interest, select a tensile test, not a compression test. However, the high susceptibility of the cake to damage/change in handling and difficulties in sizing and clamping suggest the use of alternative test modes such as shearing or penetration.

The two most common variables involved in the selection of test conditions are degree of deformation and the rate of deformation. An excellent review of the difference between small-strain non-destructive and large-strain destructive (rupture) tests was given by Bourne (1979a). Bourne and Comstock (1981) studied the effect on pretzels of the degree of compression and found dramatic increases in the quantified values of hardness, cohesiveness, springiness and chewiness at compression greater than 70%.

Shama and Sherman (1973) and Sherman and Deghaidy (1978) indicated that many foods exhibit work hardening (i.e. increased resistance with increasing deformation rates) and urged that the rate of deformation should be selected on a rational basis. Fig. 5 illustrates the effects of the degree of deformation and of the crosshead speed using pound cake as an example. No difference in firmness is found between Cakes A and B tested at small deformation and a low crosshead speed. At large deformation ($>50\%$ or 10 mm deformation) and/or higher crosshead speed (i.e. 50 mm/min) Cake B becomes definitely firmer than Cake A, in agreement with sensory evaluation.

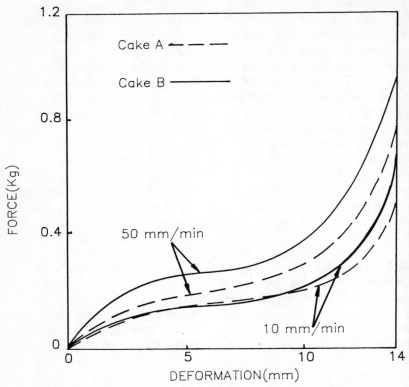

Fig. 5. Effect of the degree of deformation and Instron crosshead speed on the measured values of cake firmness (2x2x2 cm cake crumb compressed between two flat plates)

Specific TPA parameters are most meaningful measured only at certain deformation (Fig. 1). For example, when measuring cohesiveness, it is important that the sample be only partially deformed. TPA modification by using several deformations was reported by Okabe (1979) and employed by Tsuji (1978) for baked goods in Japan.

The importance of proper sample preparation prior to and during mechanical testing of baked goods can not be overemphasized. Variation in sample age, moisture content/distribution, size/shape, anisotropy and presence of crust, coating and inlays must be dealt with to assure maximum test efficiency and reproducibility. Babb (1965) demonstrated the textural variation within a loaf of bread indicating slices closer to the ends were firmer than slices taken from the center of the loaf. This suggests that the location from which the test samples are taken must be standardized.

Certain foods exhibit work softening properties (deMan 1969), therefore, the thermal/mechanical history of products prior to testing should be controlled to ascertain meaningful comparison.

Bruns and Bourne (1975) studied the snapping force of different foods and gave the following equation for size correction:

$$F = \sigma \pi R^3/L,$$

where F is the snapping force; σ, the failure stress; R, the radius of rod-shape sample; and L, the length of beam between support in a triple beam bending arrangement. Correction for sample geometry becomes important if sample comparison is to be made based on material properties. Fig. 6 illustrates this point with the snapping curves of pretzel sticks of two different radii (7.0 and 5.3 mm for Pretzel A and B, respectively). The thicker sample A showed a higher snapping force. However, its size-independent stress was lower in value, indicating that sample A is mechanically weaker.

Use of constant rate of force application (Jowitt and Mohamed, 1980) over the conventional constant deformation method appears to give cleaner, perhaps better, measurement of crispness in friable products. Fig. 7 compares two loading techniques as a piece of cracker is compressed and crushed by a 1/4" aluminum rod. It shows that crackers of varying water activity or crispness are differentiated more readily using the constant rate of force application method (A).

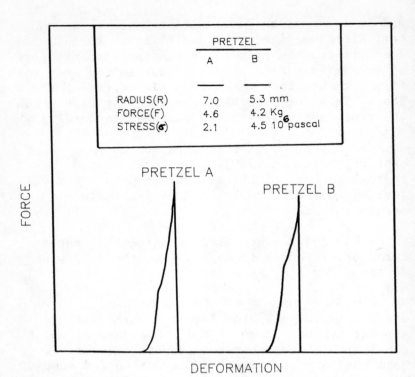

Fig. 6. Snapping curves of pretzel sticks of different size. Insert shows snapping force(F) and stress(σ) calculated according to Bruns and Bourne(1975); L=5.08 cm.

When testing soft coherent products from which crust, coating, filling and inlays (raisins, nuts, chips, etc.) cannot be removed, an increased number of replications is usually needed to obtain a meaningful average.

As soon as preliminary data are available, the differentiating ability of the test should be established and sensorially verified. The replication requirement is then determined and test conditions optimized. If multiple tests or multi-parameter test(s) were used initially, redundent tests/parameters should be eliminated.

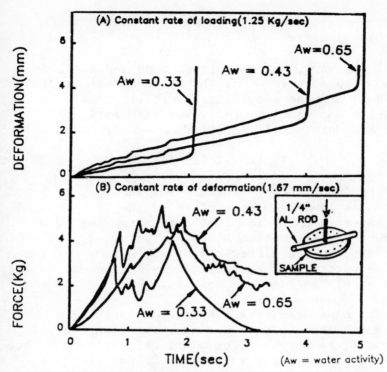

Fig. 7. Fracture properties of crackers measured as rupture deformation and rupture force under constant rate of loading(A) and constant rate of deformation(B), respectively. Insert illustrates the test arrangement.

Interpretation of rheological data on baked goods is mainly experiencial. This is particularly true for empirical tests. The basis for relating rheological data to texture acceptability or processibility is statistical correlations rather than well-thought theories. Since the values of the quantified parameters are often affected by test conditions, extrapolation from correlations obtained with different sets of conditions or samples are dangerous.

Although the kinematics of fundamental measurements are simple, the relationship of the quantified mechanical parameters to product acceptability and processability has not yet been determined. General principles are given by Ferry (1970) and Mohsenin (1970).

In summary, the test type, mode and conditions must be so selected that the obtained values will not only provide sample differentiation in a relevant and meaningful way but also serve the intended purpose efficiently and economically.

RECENT TRENDS

The importance of texture and rheology of baked goods has long been recognized by the industry and the advertising media. The increasing degree of product sophistication based on texture or textural contrast used to differentiate one's product from the competition is exemplified by the recent market success of dual-texture cookies.

The older empirical approaches, such as compression with a plunger or crushing with a blade, may still be adequate for plant QC application, but are not adequate for research. Recent trends in rheological characterization of baked goods include:

(1) Use of more descriptive imitative tests to replace the single-parameter empirical tests.
(2) Use of multiple test modes and test conditions to better simulate practical conditions.
(3) Use of universal testing machines to accomodate (1) and (2) above.
(4) Application of computer technology to facilitate test automation and the manipulation and interpretation of data.
(5) Evaluation of product in both "as is" and "wetted" forms to include important mouthfeel properties.
(6) Addition of acoustic measurement in crispness quantification of friable baked goods to complement mechanical measurements.
(7) Increasing, but limited, application of fundamental tests to basic research and structural analysis.

CONCLUDING REMARKS

The preceeding discussion has outlined the past and present use of rheology in textural evaluation of baked soft wheat products. Selected examples were given to elucidate the principles and potential pitfalls of test selection and the interpretation of results. Most of past work has been directed to the development of the hardware. More recent activities have employed commercially available universal testing instruments and have addressed test optimization by defining factors affecting the results.

ACKNOWLEDGEMENT

The author wishes to express gratitude to Dr. Alina S. Szczesniak for her technical input and editorial assistance in shaping this manuscript.

REFERENCE

American Association of Cereal Chemists, 1947. "Cereal Laboratory Methods" 5th. ed. p. 162.

Anonymous. 1972. "Tests Pastry Firmness Automatically". Food Eng. (11):89.

Babb, A.T.S. 1965. A Recording Instrument for Rapid Evaluation of Compressiblity of Bakery Goods. J. Sci. Fd. Agr. 16(11):670-679.

Bailey, L.H. 1930. A Simple Apparatus for Measuring the Compressibility of Baked Products. Cereal Chem. 7, 340-345.

Bailey, L.H. 1934. Automatic Shortometer Cereal Chem. 11:160.

Bechtel, W.G. and Kulp, K. 1960. Freezing, Defrosting and Frozen Preservation of Cake Doughnuts and Yeast Raised Doughnuts. Food Technol. 14:391-394.

Bourne, M.C. 1968. Texture Profile of Ripening Pears. J. Food Sci. 33:223-226.

Bourne, M.C. 1979a. Rupture Tests vs. Small Strain Tests in Predicting Consumer Response to Texture. Food Technol. 33(10):67-70.

Bourne, M.C. 1979b. Theory and Application of the Puncture Test in Food Texture Measurement. In "Food Texture and Rheology" (P. Sherman, ed.) pp. 95-142. Academic Press, New York/London.

Bourne, M.C. and Comstock, S.H. 1981. Effect of Degree of Compression on Texture Profile Parameters. J. Texture Science 12(2):201-216.

Brandt, M.A., Skinner, E.Z. and Coleman, J.A. 1962. Texture Profile Method. J. Food Sci. 28(4): 404-409.

Briant, A.M. and Snow, P.R. 1957. Freezer Storage of Pie Shells. J. Am. Dietet. Assoc. 33:796.

Brown, S.L. and Zabik, M.E. 1967. Effect of Heat Treatments on the Physical and Functional Properties of Liquid and Spray-Dried Egg Albumin. Food Technol. 21:87-92.

Bruns, A.J. and Bourne, M.C. 1975. Effects of Sample Dimensions on the Snapping Force of Crisp Foods. Experimental Verification of a Mathematical Model. J. Texture Stud. 6:445-458.

Choishner, Kh.D., Brindrikh, U., Puchkova, L.I. and Tarasova, L.P. 1983. Texture of Wafers and Husks. Khlebopekarnaya i Konditerskaya Promyshlennost. (7):27-28.

Civille, G.V. and Liska, I.H. 1975. Modifications and Applications to Foods of General Foods Sensory Profile Technique. J. Texture Stud. (6):19-31.

Davis, C.E. 1921. Shortening: Its Definition and Measurement. Ind. Eng. Chem. 13:797.

deMan, J.M. 1969. Effect of Mechanical Treatment on the Hardness of Margarine and Butter; J. Texture Stud. (1):109-113.

Drake, B.K. 1963. Food Crushing Sounds. An Introductory Study. J. Food Sci. 28:233-241.

Drake, B.K. 1965. Food Crushing Sounds: Comparison of Objective and Subjective Data. J. Food Sci. 30: 556-559.

Ferry, J.D. 1970. Viscoelastic Properties of Polymers. John Wiley & Son, Inc. N.Y.

Fisher, J.D. 1933. Shortening Value of Plastic Fats Ind. Eng. Chem. 13:797.

Friedman, H.H., Whitney, J.E. and Szczesniak, A.S. 1963. The Texturometer-A New Instrument for Objective Texture Measurement. J. Food Sci. 28: 390-396.

Funk, K., Zabik, M.E. and Egidsily, D.A. 1969. Objective Measurement for Baked Products. J. of Home Econ. 61:119-23.

Gaines, C.S. 1962. Technique for Objectively Measuring a Relationship Between Flour Chlorination and Cake Crumb Stickiness. Cereal Chem. 59(2):149-150.

Gaines, C.S. and Donelson, J.R. 1962. Contribution of Chlorinated Flour Fractions to Cake Crumb Stickiness. Cereal Chem. 59(5):378-380.

Guy, R.C.E. and Wren, J.J. 1968. A Method for Measuring the Firmness of Cell Wall Material of Bread. Chem. and Ind. 1727.

Gruber, S.M. and Zabik, M.E. 1966. Comparison of Sensory Evaluation and Shear Press Measurements of Butter Cakes. Food Technol. 20:968.

Harvey, A.W. 1937. Shortening Properties of Plastic Fats. Ind. Eng. Chem. 29:1155.

Heiss, R. and Witzel, H. 1969. Objective Methods for Measurement of Consistency in Solid Foods. Z. Lebensm. Untersuch u.-Forsch. 141(2):87-102.

Hirahara, S. and Simpson, J.I. 1961. Microscopic Appearance of Gluten in Pastry Dough and Its Relation to the Tenderness of Baked Pastry. J. Home Econ. 53:681.

Hornstein, L.R., King, F.B. and Benedict, F. 1943. Comparative Shortening Value of Some Commercial Fats. Food Res. 8:1.

Hutchinson, P.E. Baiocchi, F. and Del Vecchio, A.J. 1977. Effect of Emulsifiers on the Texture of Cookies. J. of Food Sci. 42(2):399-401.

Jowitt, R. and Mohamed, A.A.A. 1980. An Improved Instrument for Studying Crispness in Foods. In Food Process Engineering. Vol. 1. Link, P., Malkki, Y., Olkku, J. and Tarinkari, J. (Eds.) Applied Science Publisher, London.

Katz, E.E. and Labuza, T.P. 1981. Effect of Water Activity on the Sensory Crispness and Mechanical Deformation of Snack Food Products. J. Food Sci. 46(2):403-409.

Lowe, B., Nelson, P.M. and Buchanan, J.H. 1938. The Physical and Chemical Characteristics of Lard and Other Fats in Relation to Their Culinary Value. I. Shortening Value in Pastry and Cookie. Iowa Exp. Stal. Bull. 242.

Matthews, R.H. and Dawson, E.H. 1963. Performance of Fats and Oils in Pastry and Biscuits. Cereal Chem. 40:291.

Mohamed, A.A.A., Jowitt, R. and Brennan, J.G. 1982. Instrumental and Sensory Evaluation of Crispness. I. In Friable Foods. J. of Food Eng. 1. (1):55-75.

Mohesenin, N.N. 1970. Physical Properties of Plant and Animal Materials. Gordon and Breach Sci. Publishers. N.Y.

Morandini, W., Engle, H. and Wasserman, L. 1972. Measurement of Crumb Strength of Madeira Cakes and Yeast-Raised Baking Products, Getreide, Mehl und Brot. 26(3):68-75.

Okabe, M. 1979. Texture Measurement of Cooked Rice and Its Relationship to the Eating Quality. J. Texture Science (1):38.

Pence, J.W. and Standridge, N.N. 1958. Effects of Storage Temperature on Firming of Cake Crumb. Cereal Chem. 35:57-65.

Platt, W. and Kratz, P.D. 1933. Measuring and Recording Some Characteristics of Test Sponge Cakes. Cereal Chem. (10):73.

Platt, W. 1930. Staling of Bread; Cereal Chem. 7. 1.

Platt, W. and Powers, R. 1940. Compressibility of Bread Crumb, Cereal Chem. 17:604.

Rammell, C.G., Joerin, M.M., Craft, C.P. 1971. Quality Control Analysis of Milk Biscuits. New Zealand J. of Dairy Sci. and Technol. 6(2):57-60.

Robinson, K.S. 1971. Instrument Measures Food Texture. Food Processing Ind. (7):36-37.

Robson, A.H. 1966. The Measurement of Cake Crumb Strength. Jo Food Technol. 1:291.

Schaller, A. and Mohr, E. 1976. An Instrumental Method for Texture Measurement of Filled Waffles. L. und E. 29 (1):2-3.

Shama, F. and Shaman, P. 1968. In "Rheology and Texture of Food Stuffs", SCI Monograph. No. 27. p.77.

Sherman, P. 1970. In "Industrial Rheology". Academic Press, N.Y. p.33.

Sherman, P. and Deghaidy. 1978. Force-Deformation Conditions Associated with the Evaluation of Brittleness and Crispness in Selected Foods. J. Texture Studies (9):437-459.

Smejkalova, Z. 1974. Measurement of Rheological Properties of Bakery Products. Mlynsko-Pekarensky Primysl 20 (6):174-178.

Stinson, C.G. and Huck, M.B. 1969. A Comparison of Force Methods for Pastry Tenderness Evaluation". J. Food Sci. (34):537-539.

Szczesniak, A.S. 1963a. Classification of Textural Characteristics. J. Food Sci. 28:385-389.

Szczesniak, A.S. 1963b. Objective Measurements of Food Texture. J. Food Sci. 28:410-420.

Szczesniak, A.S. 1966. Texture Measurements. Food Technol. 20(10):52-58.

Szczesniak, A.S. 1969. The Whys and Whats of Objective Texture Measurements. Con. Inst. Food Technol. J. 2. (4):150.

Szczesniak, A.S. 1971. Consumer Awareness of Texture and of Other Food Attributes. J. Texture Stud. 2: 196-206.

Szczesniak, A.S. 1972. Consumer Awareness of and Attitudes to Food Texture. II. Children and Teenagers. J. Texture Stud. 3:206-217.

Szczesniak, A.S. 1973. In "Texture Measurements of Foods." ed. by Kramer, A. and Szczesniak, A., D. Reidel Publishing Co., Boston, Mass.

Szczesniak, A.S. and Kahn, E.L. 1971. Consumer Awareness of and Attitudes to Food Texture. I. Adults. J. Texture Stud. 2:280-295.

Szczesniak, A.S. and Kleyn, D.H. 1963. Consumer Awareness of Texture and Other Foods Attributes. Food Technol. 17:74-77.

Szczesniak, A.S. and Torgeson, K. 1965. Methods of Meat Texture Measurement Viewed from the Background of Factors Affecting Tenderness. Adv. Food Res. 14: 53-165.

Szczesniak, A.S., Brandt, M.H. and Friedman, H.H. 1963. Development of Standard Rating Scales for Mechanical Parameters of Texture and Correlation Between the Objective and Sensory Methods of Texture Evaluation. J. Food Sci. 28:397-403.

Tanaka, M. 1975. General Foods Texturometer Applications to Food Texture Research in Japan. J. Texture Stud. 6:101-116.

Toda, J., Wada, T. and Fujisawa, K. 1973b. Fundamental Textural Characteristics of Porous Rigid Foods. J. Agr. Chem. Soc. Japan. (47):513.

Tsuji, S. 1973. Method of Rating on the Texture of Baked Products by Multi-Point Mensuration Method. Intl. Cong. of Food Sci. and Technol. Abstract p. 163.

Van den Tempel, M. 1980. In "Rheometry:Industrial Applications". ed. by K. Walter. Research Studies Press, N.Y., p. 179.

Vovan, X., Castaigne, M.; Jobin, M. and Boudreau, A. 1982. Texture Measurements of Layer Cake by Different Evaluation Methods. Sci. des. Aliments 2: 195-206.

Wade, P. 1968. In "Rheology and Texture of Foodstuffs". SCI Monograph (27):101.

Yoshikawa, S., Nishimaru, S., Tashiro, T. and Yoshida, M. 1970a. Collection and Classification of Words for Description and Food Texture. I. Collection of Words. J. Texture Stud. 1:437-442.

Yoshikawa, S., Nishimaru, S., Tashiro, T. and Yoshida, M. 1970b. Collection and Classification of Words for Description of Food Texture. II. Texture Profiles. J. Texture Stud. 1:443-451.

Zabik, M.E. Fierke, S.G. and Bristol, D.K. 1979. Humidity Effects on Textural Characterization of Sugar Snap Cookies. Cereal Chem. (56):29.

RHEOLOGY OF DURUM

J.W. Dick

Department of Cereal Science and Food Technology
Fargo, ND 58105

INTRODUCTION

The rheological characteristics of durum wheat have been used by plant breeders for many years to evaluate early-generation durum wheat selections for the end-use potential in paste-type products such as pasta. The main tools used to predict durum quality in North America have been the mixograph and farinograph, although in recent years other methods such as sodium dodecyl sulfate (SDS) sedimentation, electrophoresis, and gluten measuring devices have been utilized. Also, there have been reports published indicating that the extensigraph, alveograph and amylograph are useful in predicting the quality of paste-products which may or may not be made with durum wheat.

In addition to prediction tests, various instruments and methods have also been devised which have proven useful in objectively measuring pasta cooking quality. These procedures generally involve testing the cooked pasta for textural parameters of firmness or tenderness, compressibility, elasticity or stickiness by applying forces of cutting or shearing, compression, stretching, or a combination of these to the cooked pasta.

The purpose of this paper is to briefly review some of the tests used to predict durum wheat and pasta quality in light of their usefulness to plant breeding teams and users and consumers of durum.

MIXOGRAPH

The mixograph measures the rate of dough development, resistance of the dough to mixing and the tolerance of the dough to extended mixing. The Cereal Science and Food Technology Department at North Dakota State University (NDSU) uses a 10 g mixograph procedure in its durum improvement program to evaluate relative gluten mixing strength of experimental lines. The method was developed based on a report by Finney and Shogren (1972) who described the use of the 10 g mixograph for determining and predicting functional properties of wheat flours.

The method used for NDSU breeder samples requires 10 g (as is moisture basis (mb)) of semolina and 5.8 ml of water which are mixed at an operation temperature of 24-26°C and a spring setting of eight. The mixogram from a given sample is compared to reference or standard mixograms (Figure 1) and assigned one of

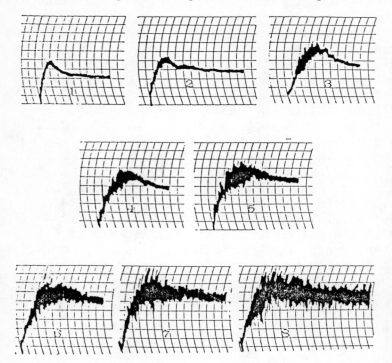

Figure 1 Reference mixograms for durum semolina at constant absorption.

eight numerical scores depending on the general mixing pattern, the higher the score the stronger the mixing characteristics. The reference mixograms were developed using durum semolina samples showing a wide range of mixing patterns.

The NDSU procedure works well for categorizing plant breeder experimental lines according to relative mixing strength and enables the breeder to select for that trait. A study by Dick and Quick (1983) of eight diverse durum wheat cultivars grown at six different locations showed that the mixogram score accounted for slightly more of the variation (36.5%) for spaghetti cooked firmness than did total wheat protein content (31.6%), but when both factors were considered together, they accounted for 64.9% of the variability in cooked firmness. Therefore, use of the mixogram score in predicting spaghetti cooked firmness is more beneficial when protein content is also considered. In a separate study using a chromosome substitution technique, Joppa et al (1984) show that durum substitution lines having high mixogram scores not only gave high spaghetti cooked firmness values but also gave high bread loaf volumes relative to those lines exhibiting low mixogram scores.

Recently, Oh et al (1985b) used the 10 g mixograph to determine the optimum water absorption of flour to prepare oriental noodles. Abrupt changes in the width and height of the mixogram upon incremental addition of water were used to predict noodle dough absorption.

The 10 g mixograph procedure when compared to the conventional 50 g farinograph procedure has the advantages of requiring a smaller sample size and allowing more samples to be tested in a given period of time. These advantages can be significant in a plant breeding situation where many samples need to be tested and the sample size is limiting. The mixograms can be assigned general mixing scores or they can be evaluated according to measurements for peak time, area under the curve, peak height or several other measurements as described by Shuey (1974). Disadvantages of the mixograph compared to the farinograph are its poor temperature control and the fact that it is not a standardized instrument, so interlaboratory results can vary considerably.

FARINOGRAPH

Probably the most important parameters that a farinograph measures include the peak time, stability and mixing tolerance index (MTI). In addition, one can obtain water absorption of a dough at a fixed mixing consistency (D'Appolonia and Kunerth, 1984). Examples of farinograms for durum and other U.S. wheat classes are shown in Figure 2.

A modified farinograph technique (Irvine et al 1961, Matsuo and Irvine 1975) has been used extensively in a plant breeding program at the Grain Research Laboratory in Winnipeg, Canada, whereby the 50 g bowl is used with the 300 g bowl linkage setting in order to measure dough mixing characteristics of durum semolina at an absorption range (26.5 to 36.0%) closer to that used to process pasta commercially. The measurements normally taken from the farinogram include the parameters:

Dough development time (DDT). The time (min) required for the curve to reach a peak.

Maximum consistency (MC). The dough viscosity in Brabender Units (B.U.) at the peak.

Tolerance Index (TI). The difference in consistency between the peak value and 4 min past the peak.

Semolina protein content influences all three parameters when samples are tested at a constant absorption as shown in Figure 3. As protein content increased, DDT decreased and MC and TI increased.

Dexter and Matsuo (1980), in a study using the modified technique of Irvine et al (1961), showed that farinograph bandwidth is a better indicator of spaghetti cooking quality than either DDT, MC or TI because it was not influenced as much by protein content as the other parameters. Using the protein

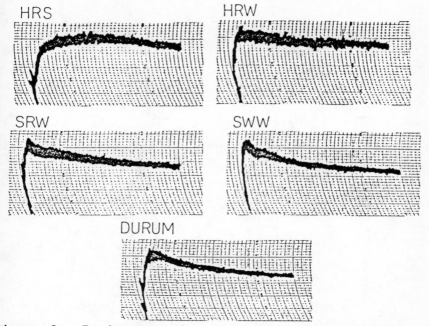

Figure 2. Farinograms for U.S. wheat classes hard red spring (HRS), hard red winter (HRW), soft red winter (SRW), soft white winter (SWW) and durum at optimum farinograph absorption.

solubility fractionation method of Chen and Bushuk (1970), they also concluded that the acetic acid insoluble residue fraction of protein was most responsible for variations in gluten strength, farinograph properties and spaghetti cooking quality.

In their study, Irvine et al (1961) showed farinogram curves (Figure 4) for two durum cultivars with different mixing strengths. A strong gluten cultivar (Pelissier) gave a relatively long DDT and a low TI, while the opposite was true for a weak gluten cultivar (Mindum). Variations in dough temperature, dough absorption, semolina particle size and semolina extraction rate (Dexter and Matsuo 1978b) were also shown to influence farinogram characteristics.

Grzybowski and Donnelly (1979) studied a series of durum wheat cultivars varying in gluten strength and protein content (12.0 to 17.6%) for their effect on spaghetti cooking quality. Relative gluten

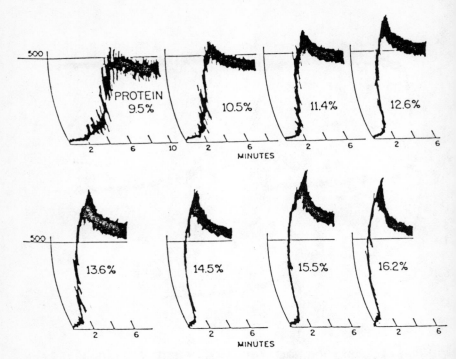

Figure 3. Effect of protein content on farinogram characteristics at 31.5% absorption (from Irvine et al 1961).

strength was determined using a conventional farinogram procedure unlike that described above. Farinograms were scored by comparison with standard reference curves (Figure 5). Cooking quality was determined by the method of Walsh and Gilles (1971). These workers (Grzybowski and Donnelly 1979) concluded that protein quantity and quality both contributed to cooking quality, and that high protein content did not necessarily guarantee optimum cooking quality. Durum cultivars with strong gluten produced spaghetti with greater cooking stability than weak gluten types as evidenced by greater retention of spaghetti firmness on extended cooking. Other workers (Dexter and Matsuo 1977, Matsuo et al 1972) testing wheats of the same variety at differing protein levels established that increased protein content improved spaghetti cooking quality.

Research by Matsuo and Irvine (1970) to study

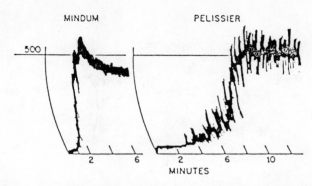

Figure 4. Farinograms representing weak (Mindum) and strong (Pelissier) gluten cultivars at 33.0% absorption (from Irvine et al 1961).

the effect of gluten from durum and non-durum wheat cultivars on spaghetti quality showed that the tenderness index of cooked spaghetti was not significantly correlated with farinogram DDT even though large differences were noted in the farinogram patterns. Gluten of medium strength appeared to produce spaghetti of optimum quality. Reconstitution studies indicated that gluten quality was the major factor determining cooking quality.

The influence of glutenin:gliadin ratios of durum wheat gluten on spaghetti cooking quality was established by separate groups (Walsh and Gilles 1971, Wasik and Bushuk 1975). Separation of proteins by gel-filtration chromatography showed that durum varieties containing high glutenin:gliadin ratios appeared to have better cooking quality. Wasik and Bushuk (1975) obtained highly significant correlations between glutenin peak area and farinograph mixing tolerance index (MTI) (-0.661), gluten strength (0.845) and tenderness index (-0.681) of cooked spaghetti.

Though desirable farinogram characteristics appear to be a prerequisite, they do not necessarily guarantee superior spaghetti cooking quality. That was the conclusion of Dexter and Matsuo (1978a) who fractionated durum and hard red spring wheat gluten proteins into six fractions (F1-F6) by precipitation from 0.005 N lactic acid at various pH levels.

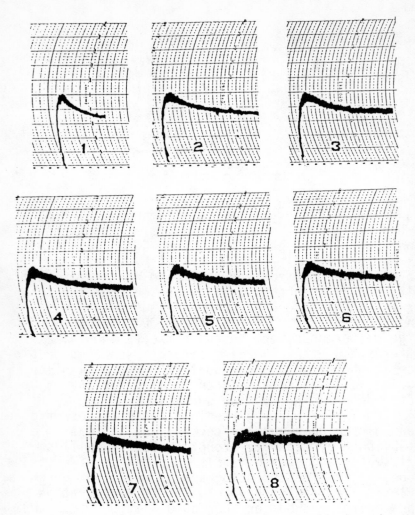

Figure 5. Standard reference farinograms for durum semolina at optimum farinograph absorption (from Grzybowski and Donnelly 1979).

Reconstitution studies revealed qualitative differences between the F4 fractions of the two wheats in their ability to improve spaghetti cooking quality. However, these differences were not reflected in a comcomitant change in farinogram mixing time.
    Dexter and Matsuo (1979) later reported on the effect of water content on changes in semolina proteins during mixing of the dough. Results of the

study showed that at both 30 and 50°C, farinograph mixing time reached a minimum value around 45% absorption (dough water content 40%) and increased when absorption was raised or lowered from that level. Data from dough protein extractability in dilute acetic acid, Osborne protein solubility fractionations, and gel filtration elution profiles of acetic acid extracts suggested that gluten development during dough mixing is limited at absorptions below 45% absorption, absorption levels at which most pasta is processed. Strong gluten wheats achieved maximum gluten breakdown during mixing at higher dough water contents than did weak gluten wheats. Matsuo et al (1982) suggested that mixing properties at relatively low absorptions, such as farinograph doughs at 31.5%, are useful for predicting extrusion properties. They felt that rheological properties at higher absorption where gluten is more fully developed might better predict textural properties of cooked spaghetti.

PROTEIN CHARACTERIZATION

In addition to the mixograph and farinograph methods, many other tests have been developed and tried in order to more precisely predict pasta quality by characterizing durum wheat protein. These tests include physical measurements of washed gluten, sedimentation rates of wheat wholemeal or flour, and solubility, electrophoresis, chromatography or specific absorption of durum wheat protein. Although most are not classified as rheological tests, they have all been correlated either with mixograms, farinograms or pasta cooking quality of durum wheat.

A gluten stretching method (Kosmina 1936) was modified (Matsuo and Irvine 1970) and a testing device developed (Matsuo 1978) to measure the force required to break a strand of wet gluten. Feillet et al (1977) described the use of a viscoelastograph to determine swelling, compressibility and recovery of a cooked disk of gluten. Other workers have reported high correlations between pasta cooking quality parameters and alveograph measurements of semolina gluten (Matsuo and Irvine 1970) and semolina dough (Walle and Trentesaux 1980).

Sodium dodecyl sulfate (SDS) sedimentation tests (Dexter et al 1980, Quick and Donnelly 1980) based on a method reported by McDermott and Redman (1977) and Axford et al (1978) compared durum wheat wholemeal sedimentation volumes to their semolina mixogram results. Both studies concluded that the SDS sedimentation test is useful for comparing gluten strengths of durum wheat cultivars when only 5-6 g of test material are available. Dick and Quick (1983) later modified the above procedures to use 1 g of wheat wholemeal or reground semolina to give a rapid estimation of gluten strength in early-generation durum wheat breeding lines. Their data (Table I) indicated that the microsedimentation (MST) procedure in combination with total wheat protein content accounted for about 71% of the variation in spaghetti cooked firmness for eight durum wheat cultivars grown at six locations and having a protein range of 11.7 to 18.3%.

TABLE I
Maximum r^2 Improvement Regression Analysis for Spaghetti Cooked Firmness (Dependent Variable)[a]

Number in Model	Variable in Model[b]	r^2
1	WP	0.316
1	MX	0.365
1	MST	0.532
2	MST, MX	0.559
2	MX, WP	0.649
2	MST, WP	0.713
3	MST, MX, WP	0.713

[a]Data from eight durum wheat cultivars grown at six locations

[b]WP = wheat protein; MST = microsedimentation value (mm); MX = 10 g mixogram score

Durum wheat proteins have been characterized by solubility fractionations (Chen and Bushuk 1970, Dexter and Matsuo 1978a, 1980), gel filtration chromatography (Walsh and Gilles 1971, Wasik and Bushuk 1975, Dexter and Matsuo 1980), reversed-phase high-performance liquid chromatography (RP-HPLC) (Bietz et al 1984), specific absorption (Dexter and Matsuo 1980, Matsuo et al 1982) using the aqueous urea method of Pomeranz (1965), and electrophoresis (Bushuk and Zillman 1978, Damidaux et al 1978, Zillman and Bushuk 1979, Damidaux et al 1980, du Cros et al 1982, Khan 1982, Khan et al 1983, Cottenet et al 1983 and Khan 1984). Generally speaking, durum proteins with greater amounts of insoluble residue protein, high glutenin:gliadin ratios and low specific absorbance showed more gluten strength on mixing and had superior pasta cooking quality. The references listed above for RP-HPLC and electrophoresis have been used to specifically characterize durum wheat gliadin proteins to identify cultivars with respect to gluten strength and cooking quality. Joppa et al (1983, 1984), using polyacrylamide gel electrophoresis (PAGE), have shown that PAGE band 45, which is associated with strong gluten, is genetically linked to chromosome 1B. Substitution of chromosome 1B from a strong gluten durum cultivar into a weak gluten durum cultivar significantly strengthened the gluten characteristics of the latter as measured by mixograph, cooked spaghetti firmness and bread loaf volume tests.

Statistical evaluation of tests to assess spaghetti-making quality of durum wheat by Matsuo et al (1982) revealed that gluten quality, farinograph bandwidth, and mixograph mixing time were significantly correlated to spaghetti cooking quality. Semolina protein content, specific absorption, SDS-sedimentation volume and farinograph bandwidth were the best indicators of cooking quality based on stepwise regression analysis.

OTHER TESTS

The extensigraph and amylograph are not commonly used to measure durum wheat quality, although some research has been done using these instruments. Grzybowski and Donnelly (1977) reported that starch gelatinization during cooking was more rapid in spaghetti containing relatively low protein levels. Dalbon et al (1982) in a study of characteristics of durum wheat starch and cooking quality of pasta, suggested that gelatinization of starch at too low a temperature and too fast swelling rate imparts a negative influence on protein network formation during cooking thus giving poor cooking quality. Relatively large starch granules which were present in the inner part of the durum grain kernel gelatinized at lower temperatures and at a faster rate than smaller starch granules as measured by the amylograph. Protein network formation during cooking was associated with both quality and quantity of gluten in the pasta. These two reports indicate that there is a competition for the cooking water between the carbohydrate and protein components of the pasta during cooking. Oda et al (1980), using the amylograph to test flour quality for use in Japanese noodles, showed that durum flour was not as acceptable for making Udon noodles as flour from soft wheat.

The extensigraph and amylograph have been used to evaluate wheat flour (non-durum) for use in sheeted Chinese noodles. Moss (1982) reported that flour exhibiting rapid starch gelatinization or with high starch paste viscosity in the amylograph gave inferior noodle quality. Physical strength in noodle doughs below a maximum resistance of 300 Brabender Units (B.U.) in the extensigraph test was also associated with inferior noodle quality. Examples of extensigrams for U.S. wheat classes are given in Figure 6.

PASTA COOKING QUALITY

Rheological measures of durum wheat are useful tools to predict pasta quality and, therefore, are valuable to plant breeding programs. Final judgement of pasta quality, however, is best done by testing the pasta itself either subjectively by human taste panels or objectively by instrumental methods. Because of the relatively small sample sizes

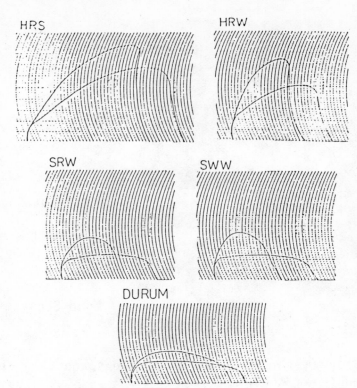

Figure 6. Extensigrams for U.S. wheat classes hard red spring (HRS), hard red winter (HRW), soft red winter (SRW), soft white winter (SWW) and durum.

available early in the plant breeding process and for increased precision, objective instrumentation procedures have become the methods of choice.

Methods using mechanical apparatus to measure cooked spaghetti firmness or tenderness have been reported by several workers (Matsuo and Irvine 1969, Walsh 1971, Voisey et al 1978a) to correlate well with taste panel results. Matsuo and Irvine (1969) and Walsh (1971) described methods for measuring the maximum force or total work required for a cutting blade to shear crosswise through a single strand of cooked spaghetti as shown in Figure 7. Measuring the forces required to cut through the spaghetti were accomplished by placing the cooked spaghetti on the surface of a load sensing cell, and while cutting the spaghetti, automatically recording

Figure 7. A single strand of cooked spaghetti being sheared (from Walsh 1971).

the force vs. distance (Figure 8) with a strip chart recorder. Voisey et al (1978a) used the same principle as the previous workers to measure cooked spaghetti firmness but tested multiple strands of spaghetti in a sample cell containing ten cutting blades.

Matsuo and Irvine (1971) improved their procedure for evaluating cooked spaghetti by measuring not only for tenderness, but also for compressibility and recovery to account for relative elasticity of the cooked spaghetti. Several years later, Oh et al (1983) described a similar technique for measuring the textural characteristics of cooked noodles. These same workers in another report (OH et al 1985a) described a method for measuring surface firmness of cooked noodles. Figure 9 represents a curve recording obtained by compressing a strand of spaghetti with a blunt compressing blade at a constant force (1.2 kg per cm^2) for a constant time (15 sec), then releasing the force and allowing the

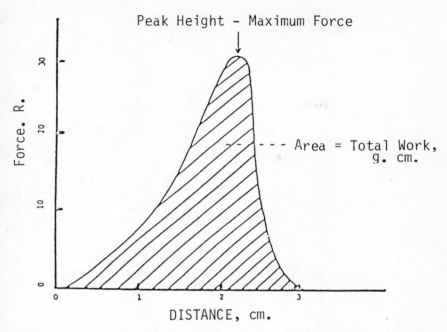

Figure 8. Force versus distance curve which shows the shearing characteristics of a single strand of cooked spaghetti (from Walsh 1971).

spaghetti to seek its resting thickness. Compressibility is defined as the ratio of y to x, and recovery is defined as the ratio of the distance that the blade is forced back (y - z) to the penetration of the blade (y). Figure 10 represents compression curves for four cooked spaghetti samples including one with a soft and mushy (Sample A) texture and one with excellent cooking quality (Sample D).

Measurement of cooked spaghetti stickiness was described by Voisey et al (1978b). The procedure used an automated device to compress ten strands of spaghetti between two flat aluminum plates, one of which was serrated. Compression force was applied to a preselected level at which point deformation of the spaghetti was stopped. The spaghetti was allowed to rest, then the plates were pulled apart to measure adhesiveness. Results of testing several samples indicated that a high starch to protein ratio at the

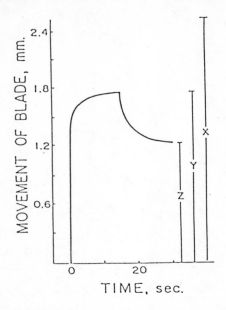

Figure 9. Values derived from compressibility curves: x = diameter of cooked spaghetti, y = blade penetration into sample; z = the extent to which the elastic component forces the blade back (from Matsuo and Irvine 1971).

surface region of the spaghetti was associated with greater stickiness.

Dexter et al (1983a) used a different automated device to measure cooked spaghetti stickiness. Using this device Dexter et al (1983b) showed that stickiness was significantly correlated to cooking loss, cooked weight, degree of swelling, compressibility, recovery and firmness, although all of these factors together accounted for less than 50% of the predicted variance in stickiness by step-up regression analysis. In a later study Dexter et al (1985) showed that the total organic matter (TOM) test of D'Egidio et al (1982), which measures the amount of material rinsed from drained, cooked spaghetti, was very closely related to instrumental measurements for cooked spaghetti stickiness, firmness and elasticity.

Data presented for evaluation of cooked samples

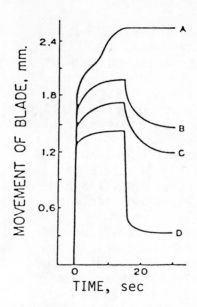

Figure 10. Compressibility curves for four cooked spaghetti samples of different quality (from Matuso and Irvine 1971).

indicate that several factors contribute to the overall eating quality of pasta. These factors include, but are not necessarily limited to, raw material ingredient quality, pasta processing conditions and the cooking water and preparation procedures.

In conclusion, while rheological testing of durum wheat gives an indication of resulting pasta quality, prediction accuracy based on these tests alone is certainly not perfect. To be most useful, rheological tests should be supplemented with other tests of quality, preferably measurement of pasta quality itself, to better ensure the end-use quality of a given wheat sample or cultivar to the commercial processor or the final consumer.

LITERATURE CITED

Axford, D.W.E., McDermott, E.E. and Redman, D.G. 1978. Small-scale tests of breadmaking quality. Milling Feed Fertil. 161(5):18.

Bietz, J.A., Burnouf, L.A., Cobb, L.A. and Wall, J.S. 1984. Wheat varietal identification and genetic analysis by reversed-phase high performance liquid chromatography. Cereal Chem. 61(2):129.

Bushuk, W. and Zillman, R.R. 1978. Wheat cultivar identification of gliadin electrophoregrams. I. Apparatus, method and nomenclature. Can. J. Plant Sci. 58:505.

Chen, C.H. and Bushuk, W. 1970. Nature of proteins in triticale and its parental species. I. Solubility characteristics and amino acid composition of endosperm protein. Can. J. Plant Sci. 50:9.

Cottenet, M., Autran, J. and Joudrier, P. 1983. Isolation and characterization of gamma-gliadins 45 and 42 compounds associated to viscoelastic characteristics of the gluten of durum wheat. C.R. Acad. Sc. Paris 297:149.

Dalbon, G., Pagani, M.A. and Resmini, P. 1982. Characteristics of durum wheat starch and cooking qualities of pasta: some preliminary considerations. Macaroni J., April:22.

Damidaux, R., Autran, J.C., Grignac, P. and Feillet, P. 1978. Relation applicable en selection entre l'electrophoregramme des gliadins et les proprietes viscoelastiques du gluten de triticum durum. Desf. C.R. Hebd. Seances Acad. Sci. Ser. D 287:701.

Damidaux, R., Autran, J.C. and Feillet, P. 1980. Gliadin electrophoregrams and measurements of gluten viscoelasticity in durum wheats. Cereal Foods World 25(12):754.

D'Appolonia, B.L. and Kunerth, W.H., editors. 1984. The Farinograph Handbook, American Association of Cereal Chemists, St. Paul, MN.

D'Egidio, M.G., De Stefanis, E., Fortini, S., Galterio, G., Nardi, S., Sgrulletta, D. and Bozzini, A. 1982. Standardization of cooking quality analysis in macaroni and pasta products. Cereal Foods World 27(8):367.

Dexter, J.E., Kilborn, R.H., Morgan, B.C. and Matsuo, R.R. 1983a. Grain Research laboratory compression tester: Instrumental measurement of cooked spaghetti stickiness. Cereal Chem. 60(2):139.

Dexter, J.E. and Matsuo, R.R. 1977. Influence of protein content on some durum wheat quality parameters. Can. J. Plant Sci. 57:717.

Dexter, J.E. and Matsuo, R.R. 1978a. The effect of gluten protein fractions on pasta dough rheology and spaghetti-making quality. Cereal Chem. 55(1):44.

Dexter, J.E. and Matsuo, R.R. 1978b. Effect of semolina extraction rate on semolina characteristics and spaghetti quality. Cereal Chem. 55(6):841-852.

Dexter, J.E. and Matsuo, R.R. 1979. Effect of water content on changes in semolina proteins during dough-mixing. Cereal Chem. 56(1):15.

Dexter, J.E. and Matsuo, R.R. 1980. Relationship between durum wheat protein properties and pasta dough rheology and spaghetti cooking quality. J. Agric. Food Chem. 28(5):899.

Dexter, J.E., Matsuo, R.R., Kosmolak, F.G., Leisle, D. and Marchylo, B.A. 1980. The suitability of the SDS-sedimentation test for assessing gluten strength in durum wheat. Can. J. Plant Sci. 60:25.

Dexter, J.E., Matsuo, R.R. and MacGregor, A.W. 1985. Relationship of instrumental assessment of spaghetti cooking quality to the type and the amount of material rinsed from cooked spaghetti. J. Cereal Sci. 3:39.

Dexter, J.E., Matsuo, R.R. and Morgan, B.C. 1983b. Spaghetti stickiness: Some factors influencing stickiness and relationship to other cooking quality characteristics. J. Food Sci. 48:1545.

Dick, J.W. and Quick, J.S. 1983. A modified screening test for rapid estimation of gluten strength in early-generation durum wheat breeding lines. Cereal Chem. 60(4):315.

du Cros, D.L., Wrigley, C.W. and Hare, R.A. 1982. Prediction of durum-wheat quality from gliadin protein composition. Aust. J. Agric. Res. 33:429.

Feillet, P., Abecassis, J. and Alary, R. 1977. Description d'un nouvel appareil pour mesurer les propietes viscoelastiques des produits cerealiers. Application al'appreciation de la qualite du gluten, des pates alimentaires et du riz. Bull. Ec. Nat. Super. Meun. Ind. Cereal 273:97.

Finney, K.F. and Shogren, M.D. 1972. A ten-gram mixograph for determining and predicting functional properteis of wheat flours. Bakers Digest 46(2):32.

Grzybowski, R.A. and Donnelly, B.J. 1977. Starch gelatinization in cooked spaghetti. J. Food Sci. 42(5):1304.

Grzybowski, R.A. and Donnelly, B.J. 1979. Cooking properties of spaghetti:factors affecting cooking quality. J. Agric. Food Chem. 27(2):380.

Irvine, G.N., Bradley, J.W. and Martin, G.C. 1961. A farinograph technique for macaroni doughs. Cereal Chem. 38:153.

Joppa, L.R., Josephides, C. and Youngs, V.L. 1984. Chromosomal location of genes affecting quality in durum wheat. Proc. 6th Int. Wheat Genet. Symp. Kyota, Japan, p. 297.

Joppa, L.R., Khan, K. and Williams, N.D. 1983. Chromosomal location of genes for gliadin polypeptides in durum wheat Triticum turgidum L. Theor. Appl. Genet. 64:289.

Khan, K. 1982. Polyacrylamide gel electrophoresis of wheat gluten proteins. Bakers Digest 56(5):14.

Khan, K. 1984. Detectability of different classes of gliadins by dye binding and trichloroacetic acid precipitability. Cereal Chem. 61(4):378.

Khan, K., McDonald, C.E. and Banasik, O.J. 1983. Polyacrylamide gel electrophoresis of gliadin proteins for wheat variety identification - procedural modifications and observations. Cereal Chem. 60(2):178.

Kosmina, N.P. 1936. Das problem der backfahigkeit. Verlag von Moritz Shafer:Leipsig.

Matsuo, R.R. 1978. Note on a method for testing gluten strength. Cereal Chem. 55(2):259.

Matsuo, R.R., Bradley, J.W. and Irvine, G.N. 1972. Effect of protein content on the cooking quality

of spaghetti. Cereal Chem. 49:707.
Matsuo, R.R., Dexter, J.E., Kosmolak, F.G. and Leisle, D. 1982. Statistical evaluation of tests for assessing spaghetti-making quality of durum wheat. Cereal Chem. 59(3):222.
Matsuo, R.R. and Irvine, G.N. 1969. Spaghetti tenderness testing apparatus. Cereal Chem. 46:1.
Matsuo, R.R. and Irvine, G.N. 1970. Effect of glu- Chem. 47:173.
Matsuo, R.R. and Irvine, G.N. 1971. Note on an improved apparatus for testing spaghetti tenderness. Cereal Chem. 48:554.
Matsuo, R.R. and Irvine, G.N. 1975. Rheology of durum wheat products. Cereal Chem. 52:131r.
McDermott, E.E. and Redman, D.G. 1977. Small-scale tests of breadmaking quality. FMBRA Bulletin, No. 6, p. 200.
Moss, H.J. 1982. Wheat flour quality requirements for Chinese noodle production. Proc. of Food Conference, Singapore.
Oda, M., Yasuda, Y., Okazaki, S., Yamauchi, Y. and Yokoyama, Y. 1980. A method of flour quality assessment for Japanese noodles. Cereal Chem. 57(4):253.
Oh, N.H., Seib, P.A., Deyoe, C.W. and Ward, A.B. 1983. Noodles. I. Measuring the textural characteristics of cooked noodles. Cereal Chem. 60(6):433.
Oh, N.H., Seib, P.A., Deyoe, C.W. and Ward, A.B. 1985a. Noodles. II. The surface firmness of cooked noodles from soft and hard wheat flours. Cereal Chem. accepted March 26, 1985.
Oh, N.H., Seib, P.A., Finney, K.F. and Pomeranz, Y. 1985b. Noodles. V. Determination of optimum water absorption of flour to prepare oriental noodles. Cereal Chem. accepted August 29, 1985.
Pomeranz, Y. 1965. Dispersibility of wheat proteins in aqueous urea solutions - a new parameter to evaluate breadmaking potentialities of wheat flours. J. Sci. Food Agric. 16:586.
Quick, J.S. and Donnelly, B.J. 1980. A rapid test for estimating durum wheat gluten quality. Crop Sci. 20:816.

Shuey, W.C. 1974. Practical instruments for rheological measurements on wheat products. Cereal Chem. 52:42r.

Voisey, P.W., Larmond, E. and Wasik, R.J. 1978a. Measuring the texture of cooked spaghetti. 1. Sensory and instrumental evaluation of firmness. J. Inst. Can. Sci. Technol. Aliment. 11(3):142.

Voisey, P.W., Wasik, R.J. and Loughheed, T.C. 1978b. Measuring the texture of cooked spaghetti. 2. Exploratory work on instrumental assessment of stickiness and its relationship to microstructure. J. Inst. Can. Sci. Technol. Aliment. 11(4):180.

Walle, M. and Trentesaux, E. 1980. Study of practical method, using the Chopin alveograph for evaluation of the suitability of durum wheat and semolina samples for preparation of pasta. Tecnica Molitoria 31(12):917.

Walsh, D.E. 1971. Measuring spaghetti firmness. Cereal Sci. Today 16(7):202.

Walsh, D.E. and Gilles, K.A. 1971. The influence of protein composition on spaghetti quality. Cereal Chem. 48(5):544.

Wasik, R.J. and Bushuk, W. 1975. Relation between molecular-weight distribution of endosperm proteins and spaghetti-making quality of wheats. Cereal Chem. 52(3):322.

Zillman, R.R. and Bushuk, W. 1979. Wheat cultivar identification by gliadin electrophoregrams. III. Catalogue of electrophoregram formulas of Canadian Wheat cultivars. J. Can. Plant Sci. 59:287.

RHEOLOGY OF BREAD CRUMB[1]

J. G. Ponte, Jr and J. M. Faubion[2]

[1]Contribution No. 86-143-B, Kansas Agriculture Experiment Station, Kansas State University, Manhattan, KS 66506

[2]Department of Grain Science and Industry, Kansas State University, Manhattan, KS 66506

INTRODUCTION

At first thought, it might seem inappropriate to apply a discussion of rheology to bread crumb. Texture, however, is among the most important factors in consumer acceptance of bread, and the objective measurement of texture does belong in the field of rheology, the science of the deformation and flow of matter. While food rheology is primarily concerned with forces and deformations, time and temperature effects are also important to many rheological phenomena. All these are certainly pertinent to the evaluation of bread crumb texture.

White pan bread constitutes the majority of the bread commercially produced in the United States. It represents over 30% of the total bakery foods market. Since texture is such an important quality attribute of this product, firmness expectations are critical and demanding. Bread is kept in the supermarket for up to 2 days, then may be kept for several additional days by the consumer. The development of large, regional bakeries with long-distance distribution compounds the problem. Thus, analysis and control of bread crumb texture is exceedingly important both from a basic, as well as an applied, viewpoint.

This chapter will attempt to address the manner in which bread texture is evaluated, the instruments used and measurements taken, sources of variability in texture measurements, and changes in texture from processing and storage variables.

IMPORTANCE OF THE RESEARCH

The measurement of bread crumb texture or rheology is, arguably, one of the longest standing fields of research in cereal chemistry. It dates to Boussingault's 1852 work on the role of moisture loss in firming of bread crumb. Certainly since early in the century, research on bread crumb texture has continued unabated. The pace of this work has increased in the last 30 years and has resulted in the development and promulgation of numerous test instruments and procedures. Likewise, it has resulted in a number of excellent review articles covering the subject with varying degrees of completeness (D'Appolonia and Morad, 1981, Elton, 1969, Herz, 1965, Knightly, 1977, Kulp and Ponte, 1981, Lasztity, 1980, Maga, 1975, Platt, 1930, Russell, 1979, Waldt, 1964 and Zobel, 1973).

While this great volume of research has obvious importance on a fundamental or basic level for our understanding of bread crumb texture, it also has considerable practical relevance. The mechanical properties of the crumb and, particularly, its compressibility or firmness are important factors in determining bread quality and, hence, consumer acceptability. The changes in crumb texture that occur during storage, ie firming, reduce product acceptability and, therefore, contribute to the 110 million pounds (Maga 1975) of bread that are returned annually as unsalable. Thus, research that provides insight into the change in textural properties, how to control or, if possible stop these changes is of great practical import.

INSTRUMENTATION AND TESTING METHODS

It is safe to generalize and suggest that the great majority of the methods and instruments used to objectively evaluate bread crumb texture are attempts to mimic the way in which a consumer subjectively evaluates that same parameter. That has traditionally been assumed to be by squeezing a slice, ie by compressibility. A number of researchers have confirmed that a correlation, in fact, does exist between subjective and instrumental measures of crumb texture (Bice and Geddes, 1949, Bashford and Hartung, 1976, Brady and Mayer, 1985). As with all textural measurements, the basis for the tests is the relationship between an applied force and resulting deformation. The result has been the development of a series of what are, essentially, consistometers in which bread crumb is compressed between parallel plates.

The first of what might be considered the modern instruments for this purpose was designed by Platt (1940). The Baker Compressimeter, which is approved as the testing instrument for the AACC standard method (AACC, 1980) of determining bread firming is, in fact, an improvement on the original design of Platt. Instruments of newer design such as the RHM loaf testing machine (Gates, 1976), the GRL Compression tester (Kilborn et al 1982) (Fig. 1), the Voland-Stevens LFRA Texture Analyzer (Voland Corp., Hawthorne, NY), and the various adaptations of the Instron Universal Testing Instrument (Instron Corp., Canton, MN) have expanded the possible range of testing conditions (variations in loading rate, for example) and opened the possibility of following time-deformation relationships (Lasztity, 1980). It is important to keep in mind, however, that all of these instruments use the same basic parallel plate geometry to apply uniaxial compression to a sample of bread crumb.

Within this instrumental or testing context, there have been two general approaches to obtaining raw data and expressing results. The

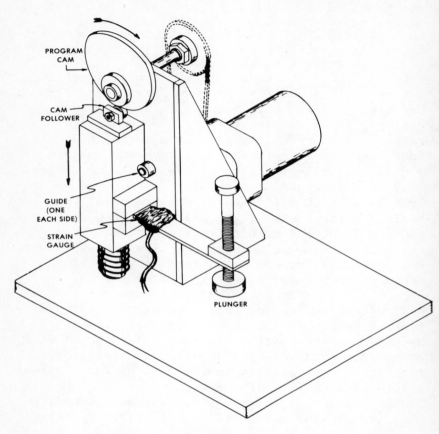

Fig. 1. Schematic diagram of the Grain Research Laboratory compression tester (from Kilborn et al, 1982).

first approach makes no attempt to express the rheological properties of the crumb in absolute physical units. Rather, it measures and expresses compressibility as one of two empirical parameters: softness or firmness. While both measures are obtained by uniaxial compression between parallel plates, they are clearly not simply reciprocals of each other. Several authors have investigated the relationship (or lack of same) between softness and firmness (Bice and Geddes, 1949, Crossland

and Favor, 1950). Perhaps it is not accidental that most of the current compression instruments can be used to measure either parameter.

Softness is measured as the deformation occurring under conditions of constant loading. When measured over a storage time of 3-4 days, softness measurements decrease in a curvelinear fashion, taking the shape of an equilateral hyperbola (Fig. 2a). Firmness, on the other hand, is measured as the force required to produce a constant deformation in the sample. When measured over the same time span, bread crumb firmness has been shown to increase in a nearly linear manner (Fig. 2b).

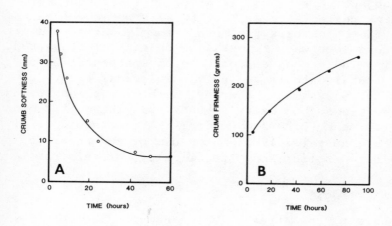

Fig. 2 (a and b). Typical "softness" (a) and "firmness" (b) curves. Softness is deformation obtained under constant load; firmness is force required to achieve constant deformation.

In attempting to distinguish between these two empirical measures of compressibility and critically evaluate each, Bice and Geddes (1949) suggested that crumb softness curves could not be readily interpreted, since their slopes were complicated functions of both rate of change with time and the original texture of the bread being tested. Firmness data, conversely, were linear plots, whose slopes were simple, direct functions of relative rates of change in firmness. The validity of their conclusion, that firmness data were a more reliable index of the textural change taking place in bread crumb over time, has been borne out by the fact that most compressibility data are expressed as measures of firmness. It is worth noting that, from a logical standpoint, measurement or evaluation of bread crumb firmness is more reliable than that of softness. Softness data suggest that most of the textural changes taking place in bread crumb occur rapidly within the first day of storage. Organoleptic or subjective evaluations of breads over that same time period suggest that this is not the case and that the changes are more strongly correlated with those seen when firmness is objectively measured.

The second approach to the presentation of textural data is to apply rheological theory to the raw data and, thereby, determine a modulus that describes the observed behavior. The most common treatment (Bice and Geddes, 1949, Cornford et al, 1964, Russell, 1983) makes use of the theory of elastic solids and, specifically, Hooke's Law;

$$\sigma = E\varepsilon$$

where σ, the stress, is the compressive force/unit area, ε, the strain, is the fractional deformation of the sample resulting from Σ and E is Young's or the Elastic modulus (expressed in dynes/cm^2).

Analysis or interpretation of data by this method can be open to a number of problems. As long ago as 1949, Bice and Geddes (1949) noted that fresh bread crumb did not obey Hooke's law and that the relationship between stress and strain in the sample changed as the bread aged.

Application of Hooke's law presumes that the material being tested is an ideally elastic body or, at least is behaving in that manner under the stress, strain, and time conditions employed in the test. Bread crumb, however, does not generally behave as an ideally elastic material. It is, in fact, viscoelastic and the range of stresses over which bread crumb behavior is linearly elastic is low, narrow, and very poorly defined (Lasztity, 1980).

The structure of bread crumb contributes to the problem of application of viscoelastic theory. Crumb is a porous matrix or lattice. When compressed, as with the Baker compressimeter, Instron, etc, the crumb experiences not only compressive stress but also flexural and shear stresses. This complex combination of stresses makes mathematical description of the deformation process as a relationship between stress and strain a very difficult task. Generally, an apparent elastic modulus is all that can or should, be calculated.

Not all techniques for measuring crumb rheology rely on parallel plate compressibility of individual slices of bread. Reasoning that the measurement of crumb cell wall firmness should be separable from effects caused by cell size and shape variation, Guy and Wren (1968) suspended bread crumb in pure hydorcarbon and centrifuged it to create an artificial pellet. Crumb texture was evaluated as firmness using the Instron to supply constant compression. Interestingly, crumb cell wall pellets appeared to recover completely after compression. This allowed the authors to retest the same pellet over a period of several days and, thereby, construct a firming curve using only 10 grams of crumb.

Willhoft (1971a) used gaseous compression of wrapped whole loaves to determine the bulk compression moduli and recoverable volumes of deformation (RDV) of the crumb. Loaf RDV, which reflects both the instantaneous elastic and retarded plastic properties of the material

(Bishop and Wren 1971), decreased curvelinearily from 1.41 to 0.41 over 7 days of storage. This corresponded to increases in bulk compression moduli of from 206 to 724 Kdyn/cm^2. Using a specially designed test cell on the Instron Universal Testing Machine, Dahle and Montgomery (1978) subjected bread slices to a combination of compression and shear to the point of visible rupture. Total deformation (in mm) before rupture was taken as a measure of extensibility. Maximum resistance (in grams) before rupture was defined as crumb strength. As might be expected, crumb extensibility decreased and strength increased over 5 days of storage. Storage at reduced equilibrium relative humidities (ERHs) increased the magnitude of both of the above changes. Brief storage (1/2 hour) of the slice at elevated temperatures increased extensibility and decreased strength. Changes in the hydration of the gluten matrix were cited as being responsible for the differences seen with both treatments.

McDermott (1974) employed an ingeniously modified tri-beam balance as part of an investigation of crumb texture of bread recently removed from the oven. Compressibility and crumb resilience were measured as deformation under a fixed load and recovery after unloading respectively. Adhesion or crumb stickiness was measured as the weight required to remove a plexiglass disc (pressed onto the crumb during loading) from the crumb surface. Crumb was stickiest at the center of the slice and the center of the loaf. Crumb resilience also varied throughout the loaf, such that the highest values occurred at the center of the loaf and near the crust. When measured at times up to 3 hr out of the oven, crumb stickiness decreased rapidly over the first hour of cooling. After that time, only small changes were seen. Crumb resilience, on the other hand, showed an opposite trend over the same time span, increasing rapidly over the first hour and slowly during hours 2 and 3. As might be anticipated, increases in flour malt levels produced crumbs of greater stickiness and lower resilience.

LIMITATIONS OF THE TECHNIQUES

Objective assessment of bread crumb texture is, by no means, straightforward or simple. As a consequence, interpretation and comparison of results are often difficult. The situation is elegantly and concisely explained by Hibberd and Parker (1985).

> "A wide range of sample sizes and shapes and many modes of deformation have been used. Also, many different parameters have been derived from the raw experimental data. Because bread crumb is not a rheologically simple material, the applied force is not linearly related to the corresponding deformation and it is not generally possible to compare results from different instruments. The results also depend on the rate of deformation because bread crumb has viscoelastic properties. Thus, it is not possible to assess and compare critically the results from many of the published papers as specific details of the experimental methods and sampling procedures are not presented and few authors report the effects of varying the conditions of measurement. Even fewer present the data in a form which indicates the reproducibility of the results, which is necesarily limited by the variations between samples as well as the errors in the measurement procedure."

Added to this is the great variability in bread formulation, production regimes, and baking procedures. Underlying all the above is the fact that differences in the inherent breadmaking quality of the flours used may obscure otherwise relevent comparisons between studies using the same instrument or measurement technique.

PROBLEMS OF BREAD CRUMB UNIFORMITY

The previous sections have shown that a number of instruments are available that will measure crumb rheological properties in a more or less satisfactory manner. Perhaps the greatest impediment to the accurate assessment of these properties is the problem of sample preparation. Bread crumb is not homogeneous, and one could readily find greater differences within a single loaf than between loaves representing different treatments.

Variability within loaves is not surprising, when one considers the sequence of events that occurs during bread baking. The moulded and proofed pieces of dough enter the oven at temperatures of about $40^{\circ}C$. Since oven temperature is typically about $217^{\circ}C$, internal dough temperatures will obviously rise, but since dough exhibits insulating properties, the temperature rise will not be uniform throughout the baking loaf.

Fig. 3, from Marston and Wannan (1976), shows that the temperature rise at the center of the loaf is clearly slower than that at the outer portions. Given this temperature gradient, one would also expect gradients of moisture content to exist as well as of yeast and enzyme activities. Varriano-Marston et al (1980), using techniques including enzymatic hydrolysis, polarization microscopy, and x-ray diffraction, showed that the extent of starch gelatinization in the center of a loaf was different from that of the outer portions. It is not surprising, therefore, that several workers have found that bread firmness is not homgeneous throughout the loaf. Ponte et al (1962) showed that the center of a loaf of bread was firmer than the end

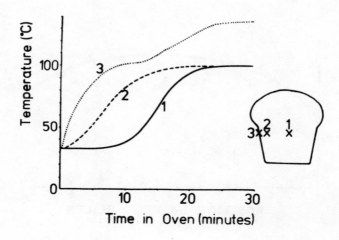

Fig. 3. Temperature-time profiles from a dough piece baked at 235°C. From Marston and Wannan (1976).

portions (Fig. 4) and suggested that variations in processing techniques may affect this pattern.

Hibberd and Parker (1985) recently studied a number of factors leading to variability in bread firming measurements. Fig. 5 illustrates some of their findings. The data, which show force as a function of crumb deformation measured in three directions (either at the end or middle portions of the loaf), demonstrate firmness differences depending on the direction of compression. Differences caused by position of the measured sample, middle or end, are also confirmed. Thus, this clearly demonstrates that a loaf of bread is not a homogeneous entity in terms of crumb rheological properties. Any firmness measurement

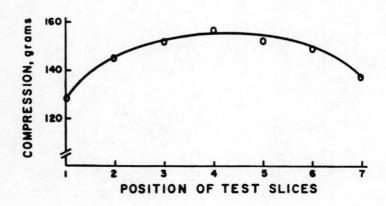

Fig. 4. Variation in crumb compressiblity within a loaf.

method must be designed to take this into account.

RHEOLOGICAL PROPERTIES RELATED TO SENSORY EVALUATIONS

Staling is well-recognized as a complex phenomenon, embodying changes in flavor and aroma, as well as textural changes in both crumb and crust. Crumb textural changes are widely employed for both research and industrial quality control purposes to measure "staling." This is the case for several reasons: textural (firmness) changes are relatively easy to measure objectively compared to flavor and aroma changes; firmness of bread (as determined by subjective squeeze tests) is an important criterion for the consumer in

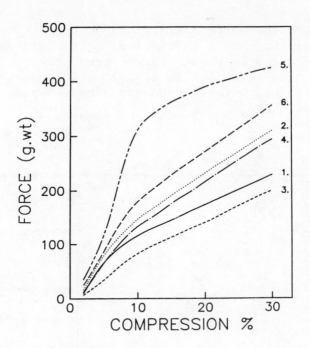

Fig. 5. Typical plots of force versus compression for the first compression cycle for "fresh" bread crumb. Compression direction: 1. Parallel to lateral axis, middle position. 2. Parallel to lateral axis, end position. 3. Parallel to vertical axis, middle position. 4. Parallel to vertical axis, end position. 5. Parallel to long axis, middle position. 6. Parallel to long axis, end position. From Hibberd and Parker (1985).

making a purchase judgement; and firmness characteristics of the crumb are, indeed, an important component of bread freshness. However, as was previously mentioned, it is obvious that objective rheological measurements are meaningful only to the extent that they correlate with a sensory perception of overall staling. Several studies have indicated that a rheological test could be a valid predictor of bread freshness.

Cornford et al (1964) successfully correlated crumb modulus with panel ratings of freshness.

Bashford and Hartung (1976) supported the possibility that a rheological test could be standardized to predict bread freshness. Fig. 6, taken from Axford et al (1968), illustrates a good relationship between crumb modulus and panel rating.

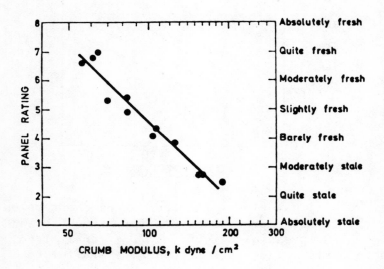

Fig. 6. Organoleptic panel rating of bread freshness correlated with crumb firming. From Axford et al (1968).

That care must be taken to achieve satisfactory relationships between rheological tests and sensory evaluations has also been demonstrated by Moskowitz et al (1979). They showed that the elastic modulus and hysteresis loss were associated with bread textural characteristics as measured by panel ratings. However, observed correlations varied with mechanical deformation, sample dimensions, and whether or not a consumer or expert panel was employed for testing. These authors suggested that further work was needed to standardize both sensory and firmness testing. Brady and Mayer (1985) attempted to relate Instron compression data to sensory evaluations of commercial rye and French loaves. These authors derived measures of hardness, cohesiveness, elasticity, and chewiness using the Instron and the Texture Profile Curve of Bourne (1978). Comparing instrumental measurements to sensory testing by a trained panel, a stronger relationship was found with rye bread than with French. The fact that significant differences were found among samples by instrumental evaluation and not by sensory testing, suggested to the authors that further work was needed to ascertain whether the two tests were measuring the same textural parameters.

Origins of Bread Crumb Structure

The classic work of Sandstedt et al (1954) provides the basis for understanding the structure of bread crumb. Dough and bread crumb sections were embedded in plastic, sectioned by grinding, and the exposed surface was stained to selectively highlight protein and the starch.

Using this technique, freshly mixed dough was demonstrated to be a disorderly or nonoriented arrangement of starch granules in a continuous matrix of protein. After fermentation but before baking, the starch granules showed some orientation with reference to the protein matrix surface.

During the baking process, gas cells expanded and exerted a pressure or force on the granules parallel to the protein film surface. As the granules became gelatinized and, thus, softened, they were dilated, elongated, and more regularly oriented with reference to the cell surface. This is shown in Fig. 7.

Dennet and Sterling (1979) also found that gelatinized wheat starch was most uniformly oriented in fibrous strands of crumb and in films about the walls of gas cells. Gelatinization of starch, however, is not complete in crumb.

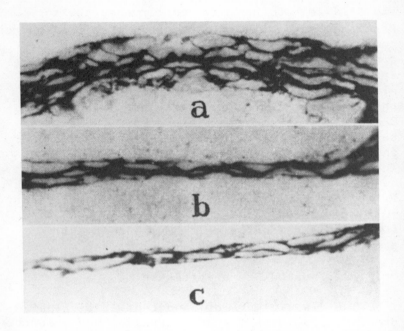

Fig. 7. Orientation and elongation of starch granules in cross sections of bread films (X200). Sandstedt et al (1954).

Marston and Wannan (1976) point out that the ratio of water to starch in dough is about 1:1, whereas a ratio in the order of 3:1 is required for complete gelatinization. Thus, this basic structure of a single gas cell, on a macro scale, represents the system that is deformed when rheological tests are applied to bread. The structure is modified depending on formulation and processing conditions used to produce a given bread. For example, French-type, hearth bread would be expected to have thicker cell walls than a typical, fine-celled, white pan bread. The crumb of sponge dough bread, subjected to the usual make-up procedure, would have a somewhat different structure than its counterpart produced by the continuous mix breadmaking method which, in the absence of dough make-up, exhibits little grain orientation.

EFFECTS OF FLOUR COMPONENTS

Considerable effort has been expended over the years to investigate the role of the principal flour components: starch, protein, pentosans, and lipids in bread firming. Much of this work and its results are summarized in the reviews cited above.

Most investigators have ascribed a central role in bread firming to the starch component of bread flour. An early worker in this area, Katz (cited in Maga, 1975), attributed firming of bread to the retrogradation of starch in the crumb. Subsequent work by Schoch and French (1947) indicated that the heat-reversible aggregation of amylopectin is primarily responsible for bread-staling. Both of these workers concluded that this aggregation was the result of intermolecular attractions between linear side-chains of branched molecules, a type of association less rigid than that involved in the retrogradation of amylose. A later report by Schoch (1965) assigned a subordinate role to amylose, indicating that this starch component associates rapidly and is already retrograded in freshly baked bread. According to Schoch's model, amylopectin associates gradually, and is primarily responsible for the firming that takes place as bread ages.

The relationship between increased starch crystallinity and firming has been studied by a number of investigators. Cornford et al (1964) reported that the relationships among elastic modulus, time, and temperature in bread crumb are consistent with a physical process. This same process was, presumably, the cause of the increase in crumb modulus involving an increase of crystallinity in the material. These workers utilized the Avrami equation, an expression derived from the kinetic theory of phase changes, to model the change in crystallinity of a system.

Kim and D'Appolonia (1977 a,b,c) also used kinetic studies and the Avrami equation, and found that the basic mechanism of staling involved changes analagous to crystallization of the starch fraction of the crumb. However, not all researchers agree that changes in crystallinity of the starch component of the crumb are the major factor affecting bread staling. Dragsdorf and Varriano-Marston (1980), using x-ray diffraction, concluded that starch crystallization and bread firming are not equivalent. Ghiasi et al (1984) utilized differential scanning calorimetry to follow changes in crystallinity of bread crumb as it aged and then was freshened by reheating. They determined that the observed change in the amount of crystallinity did not correlate well with changes in crumb firming.

The role of flour protein in the crumb firming process has also been studied for many years, but this relationship is still unclear. Investigators such as Bechtel and Meisner (1954), working with reconstituted flours of varying protein content, found that all the breads staled at the same rate during the first 3 days of storage. Later in storage, the bread produced from higher protein reconstituted flours staled less rapidly. Cluskey et al (1959) appeared to corroborate these trends by showing that the rigidity of flour and starch gels increases rapidly during the first 2 days after preparation, but relatively less firming occurs in gluten during the same period. Prentice et al (1954), who worked with reconstituted flour systems, observed that an increase in protein led to decreased firmness along with increased loaf volume. It is difficult to evaluate findings such as these, since it is known that higher specific volumes of bread are associated with decreased firmness.

Willhoft (1971a,b) explained the antifirming effect of gluten on the basis of (a) direct dilution of starch or (b) the effect of gluten enrichment on loaf volume. Kim and D'Appolonia (1977b) found that increased protein decreased bread firming, but their kinetic studies indicated that this effect was due to dilution of starch and not the quality of the protein. More recently, Maleki et al (1980) conducted firming studies using reconstituted flours and determined that the factor affecting differences in firming rates was associated with the gluten protein fraction.

Although wheat flour pentosans are present in small quantities (2-3%), they have received much attention over the years because of their unique properties. In dough, they absorb up to 10 times their weight in water (Bushuk, 1966). Also, the water-soluble constituents of flour, primarily water-soluble pentosans, form gels in the presence of oxidizing agents at room temperature. The mechanism of this phenomenon has been studied by various workers (e.g., Neukom and Markwalder, 1978; Sidhu et al, 1980). The role of pentosans in crumb rheological properties is not clear. Studies performed to determine the effects of pentosans on loaf volume potential and other quality factors have produced contradictory results, presumably because of the differing experimental techniques employed (Shelton and D'Appolonia, 1985). One of the studies aimed at understanding the effects of pentosans on staling was that of Kim and D'Appolonia (1977c), who found that added pentosans, especially the water-soluble fraction, decreased the firming rate. Kinetic studies revealed that this anti-firming effect was due to dilution of the starch components available for crystallization. Generally, it is known that the addition of pentosan gums to dough will increase the moisture of bread, and higher crumb moisture will enhance bread freshness (ie reduce firming), provided loaf volume has not been adversely affected.

Lipids are also a minor constituent of wheat flour, whose contribution to crumb rheology appears to be important, if still not completely understood. A recent review in this field

(Chung and Pomeranz, 1981) has concluded that wheat flour polar lipids are important functionally because of their effective interactions with gluten and, perhaps, with starch during baking. These same polar lipids also play a major role in the "shortening response" in breadmaking.

EFFECTS OF FORMULA INGREDIENTS

Bread crumb rheological properties are influenced to some extent by the type and amount of ingredients used in breadmaking. Many studies have been conducted over the years to delineate effects of ingredients. The results have been confusing in some cases. Complex interactions among ingredients, processing techniques, and product characteristics undoubtedly have contributed to this confusion.

A subjective compilation of ingredient effects on bread freshness is given in Table 1. Further information is available in the review of Maga (1975). The present discussion will focus attention on only two ingredient types listed in Table 1, enzymes and surfactants.

Miller et al (1953) showed that three types of alpha-amylase preparations (malted wheat flour, fungal, and bacterial) reduced the firmness of bread compared to untreated controls. Fig. 8 illustrates part of their findings. The efficacy of these amylases in reducing bread firmness was positively related to their thermostability. Bacterial amylase, possessing the highest degree of heat stability, appears to partially survive the baking process and, therefore, can affect the greatest amount of starch degradation. Fig. 8 shows that bacterial amylase substantially decreases the crumb firming rate. A number of studies have suggested that bacterial amylase is an effective anti-firming agent (see Kulp, 1975). This enzyme, however, has not been commercially successful because its activity has proved to be difficult to control. If excessive levels are

Table 1. Effect of formulation of white pan bread on its freshness.

Formula ingredient	Crust Freshness[a]	Crumb Freshness[a]
Flour protein	+	+
Sugars	+	+
Oligosaccharides	+	+
Dextrins	+	+
Milk ingredients	+	−
Milk replacers	+	±
Salt	±	±
Shortening	−	+
Water absorption		
High	−	−
Optimum	+	+
Low	−	−
Enzymes		
Malt	+	+
Fungal amylases	+	+
Bacterial amylase	+	++
Surfactants	+	++

[a] + = Improves freshness retention.
± = No effect on freshness retention.
− = Reduces freshness retention.
From Kulp, K., Am. Inst. Baking, Tech. Bull., 1(8), 1979.

used, or if bread storage temperatures are higher than anticipated, enzyme activity proceeds to a point where the crumb becomes gummy, sticky, and weak. Silberstein (1964) found that good results were obtained when he used combinations of bacterial alpha-amylase and a surfactant. Possibly, the surfactant reduced bacterial amylase variability by complexing with the enzyme. Recently, DeStefanis and Turner (1981) reported that bacterial alpha-amylase could be successfully

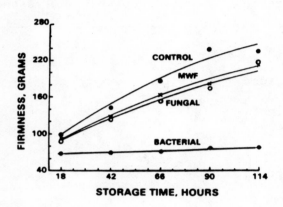

Fig. 8. The effects of equivalent levels of malted wheat flour, fungal, and bacterial alpha-amylase on bread firmness. From Miller et al (1953).

used to control bread softness if proteolytic enzymes associated with the amylase were first inactivated. Their patent indicated that the purified enzyme preparation could be used alone or in combination with approved surfactants to lower the rate of firming, without loss of product quality

Perhaps the most widely utilized means of minimizing bread firmness by the U.S. baking industry is the use of surface active agents (surfactants). A number of these compounds are presently permitted by the FDA Standards of Identity (Kulp and Ponte, 1981). Some surfactants (e.g., sodium stearoyl-2-lactylate and succinylated monoglycerides) exhibit both crumb softening and dough strengthening properties, whereas mono- and diglycerides, surfactants with the longest history of usage, perform only as crumb softeners.

It is well recognized that, in breadmaking, surfactants exert their effects through actions other than classical emulsification mechanisms. The crumb softening surfactants reduce firmness in bread via interactions with starch components, as recently reviewed by Krog and Davis (1984). D'Appolonia (1984) also reviewed studies in this area, and reported that surfactants decreased the amount of amylose found in the water-soluble fraction of the bread crumb, further confirming interaction between surfactants and starch components.

The effect of surfactant on crumb rheological properties is illustrated in Fig. 9, from Kulp and Ponte (1981). The "softener" used in this study was a commercial mono- and diglyceride. As the figure indicates, the greatest effect of the surfactant was to reduce overall firmness of the bread, i.e., to make the bread "softer" to begin with. After a storage period of 96 hr, the bread with surfactant was approximately equivalent in firmness to control breads after 48 hr of storage. Fig. 9 also shows that the surfactant slightly lowered the firming rate of bread. Whether or not surfactants lower firming rates has been the subject of some debate in past years (Kulp and Ponte, 1981).

FACTORS AFFECTING BREAD FIRMING

Aside from the effects of ingredients, a number of factors affect crumb rheological properties. A few of these effects are discussed below. The previously cited reviews provide a more detailed treatment.

Volume (not surprisingly) has an influence on crumb rheology and firming. For loaves of equivalent weight, differences in volume generally imply differences in cell wall thickness and size

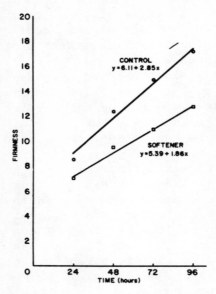

Fig. 9. Changes in rate of bread firming. Effect of a commercial surfactant (mono- and diglycerides). From Kulp and Ponte (1981).

of the cells themselves. Such differences would be expected to affect fundamental rheological properties. Also, of course, differences in weight per unit volume when testing crumb samples for firmness would be expected to influence the forces required to deform the samples. Axford et al (1968) showed that loaf specific volume was a major factor in measuring both the rate and extent of firming. Both factors decrease in a linear manner as loaf volume increases.

Temperature also influences crumb rheology. Cornford et al (1964) demonstrated that the rate of bread firming increased as the storage temperature (-1° to 32°C) was lowered. These workers noted that all breads tended to firm to the same ultimate firmness, but that firming rate was dependent on storage temperature. More recently, Cole (1985) also showed that bread stored at lower temperatures (7.5° to 25°C) firmed more rapidly than control. An interaction between storage temperature and fat was noted. Specifically, increasing shortening (hydrogenated soy oil) from 1.5 to 3.0% caused bread to firm at an increased rate at the lower storage temperature. Other fats (lard or soy oil) caused decreased firmness when used at levels of 0 to 3%, irrespective of storage temperature.

The baking industry generally recognizes the detrimental effects of decreased temperature on bread firmness, and responds by attempting, during periods of cold weather, to distribute bread products under controlled temperature conditions (e.g., heated delivery vans), insofar as possible.

Water in bread systems affects crumb rheology in several ways, as pointed out by Kulp and Ponte (1981). Water is involved in drying processes, moisture equilibraton between bread crumb and crust components, and moisture redistribution between and among basic bread components (e.g., protein, starch).

If bread loses moisture during storage, it becomes firmer and less acceptable. Bechtel et al (1953) showed that, with bread at a constant age, a sensory panel perceives bread with higher moisture levels as being fresher. Moreover, bread with higher water levels was generally found to stale at a slower rate. Therefore, it is in the interest of both the consumer and producer to maintain white bread moisture levels as close to the legal limit of 38% as is practical. However, as has been known for many years, bread will firm regardless of moisture loss.

Upon completion of baking, moisture equilibration between bread crumb and crust will continue. Moisture, of course, will diffuse from the inner portions of the loaf to the crust, which has lost substantial moisture during baking. The most recent data illustrating this phenomenon have been summarized by D'Appolonia (1984), who further observed that the presence of surfactants slows the transfer of moisture from crumb to crust. The redistribution of water between crumb and crust has a bearing on crumb firmness, as noted above. It also changes the character of the crust, causing it to become less crisp.

As bread crumb firms with age, the crumb appears to "dry out", whether or not overall moisture is actually lost. Presumably, a redistribution of moisture among bread constituents occurs, and this movement of moisture would be expected to affect crumb rheological properties. This problem has been studied for many years, with contradictory results. Cluskey et al (1959) observed that gluten did not change in water-sorption characteristics when baked by heating with excess water at $98^{\circ}C$ for 15 min. The moisture-sorption capacity of starch treated in a similar manner was increased substantially. On the basis of their studies, these workers concluded that moisture is transferred from starch to gluten in bread crumb as a function of time. Willhoft (1971a, 1972) separated gluten and starch by ultracentrifugation, and subjected these components, placed in a special diffusion cell to heating at $95^{\circ}C$ for 1 hour and cooling for another 1 hour. The moisture of the gluten and starch components was periodically analyzed. The results of these experiments were interpreted as indicating moisture movement from gluten to starch during baking and subsequent storage. This, of course, is opposite to the conclusion reached by Cluskey et al (1959). The reason for such divergence of opinions, as suggested by Kulp and Ponte (1981), may be that model-system studies, such as those above, oversimplify the conditions found in actual bread. Factors not accounted for in model systems include the degree of swelling and hydration of starch, degradation of starch by carbohydrases, level and type of saccharide ingredients, presence of wheat flour pentosans, and the effects of various lipid materials.

Another factor known to influence crumb rheology and bread firming is the production method utilized to make the bread, including the operational parameters within any given method. Sponge dough bread typically has slightly more volume and a softer crumb than straight dough bread. Both methods employ a bulk fermentation period, but they are sufficiently dissimilar that differences in gluten, starch, and/or other components affect the firming of bread crumb.

Many baking technologists agree that short-time (i.e., "no-time") dough bread is characterized by a crumb with poorer keeping qualities than sponge dough bread. Recently, Kai (1985) provided data that confirmed this observation. Short-time dough bread, whether the dough was mixed conventionally or in a high-speed mixer, exhibited a firmer crumb over an 8-day storage period than sponge dough bread. This difference in firming appeared to be related to the "condition" of the starch in the crumb, as indicated by pasting properties of the crumb measured in the Brabender viscoamylograph.

The continuous-mix dough system is known to produce bread with characteristically different crumb properties. This bread, compared to "conventional" breads, had a exceptionally uniform, fine grain, and had a crumb that was typically very soft and relatively fragile.

As suggested above, operational parameters within any given production method could also affect crumb properties. The type and extent of mixing imparted to the dough, moulding, fermentation, and baking all could exert some influence.

CONCLUSIONS

Much empirical work has been done over the years on the rheology of bread crumb. Most of this work has been directed at measuring and understanding bread crumb firming as an indicator of staling, a problem of long-standing significance to the baking industry. Present hypotheses are not generally satisfactory to explain changes in bread textural properties as a function of bread age. Newer rheological methods involving dynamic stress-strain measurements (see chapter: Dynamic Rheological Testing of Wheat Flour Doughs, in this volume) appear promising, and may lead to advancements in knowledge of crumb rheology, an area in which relatively little basic work has been done. After many years of study, the means available to control changes in crumb rheological properties, i.e. bread firmness, remain quite limited.

LITERATURE CITED

American Association of Cereal Chemists. 1980. Approved Methods of the AACC. Method 74-10, aproved April, 1961. The Association, St. Paul, MN.

Axford, D., Colwell, K., Cornford, S. and Elton, G. 1968. Effect of loaf specific volume on the rate and extent of staling in bread. J. Sci. Food Agric. 19:95.

Bashford, L. and Hartung, T. 1976. Rheological properties related to bread freshness. J. Food Sci. 41:446.

Bechtel, W., Meisner, D. and Bradley, W. 1953. The effect of crust on the staling of bread. Cereal Chem. 30:160.

Bechtel, W. and Meisner, D. 1954. Bread staling studies of bread made with flour fractions. I. Fractionation of flour and preparation of bread. Cereal Chem. 31:163.

Bice, C. and Geddes, W. 1949. Studies of bread staling. IV. Evaluation of methods for the measurement of changes which occur during bread staling. Cereal Chem. 26:440.

Boussingault, J. 1852. Experiments to determine the transformation of fresh bread into stale bread. Ann. Chim. Phys. 36:490.

Bishop, E. and J. Wren. 1971. A method for measuring the firmness of whole loaves of bread. J. Fd. Technol. 6:409.

Bourne, M. C. 1978. Texture profile analysis. Food Technol. 32(7):62.

Brady, P. and Mayer, M. 1985. Correlations of sensory and instrumental measures of bread texture. Ceral Chem. 62:70.

Bushuk, W. 1966. Distribution of water in dough and bread. Bakers Dig. 40:38.

Chung, O. K. and Pomeranz, Y. 1981. Recent research on wheat lipids. Bakers Dig. 55:38.

Cluskey, J. E., Taylor, N. W. and Senti, F. R. 1959. Relation of the rigidity of flour, starch, and gluten gels to bread staling. Cereal Chem. 36:236.

Cole, F. A. 1985. Bread Staling: Effects of fats, surfactants, storage time and storage temperature, and the interaction between these factors. M.S. Thesis, Kansas State University, Manhattan, KS.

Cornford, S., Axford, D. and Elton, G. 1964. The elastic modulus of bread crumb in linear compression in relation to staling. Cereal Chem. 41:216.

Crossland, L. B. and Favor, H. H. 1950. A study of the effects of various techniques on the measurement of firmness of bread by the Baker Compressimeter. Cereal Chem. 27:15.

Dahle, L. and E. Montgomery. 1978. A method for measuring strength and extensibility of bread crumb. Cereal Chem. 55:197.

D'Appolonia, B. L. 1984. Factors for consideration in bread staling. Chapt. T-1 in: International Symposium on Advances in Baking Science and Technology. Tsen, C. C. ed. Dept. of Grain Science and Industry, Kansas State University, Manhattan, KS.

D'Appolonia, B. and M. Morad. 1981. Bread Staling. Cereal Chem. 58:186.

Dennet, K. and Sterling, C. 1979. Role of starch in bread formation. Staerke. 31:209.

DeStefanis, V. A. and Turner, E. W. 1981. Modified enzyme system to inhibit bread firming. Method for preparing same and use of same in bread and other bakery products. U.S. Patent 4,299,848. Nov. 10, 1981.

Dragsdorf, R. D. and Varriano-Marston, E. 1980. Bread staling: X-ray diffraction studies on bread supplimented with alpha-amylase from different sources. Cereal Chem. 57:310.

Elton, G. 1969. Some quantitative aspects of bread staling. Baker's Dig. 43:24.

Gates, R. 1976. The RHM loaf testing machine. Fd. Proc. Ind. 45:30.

Ghiasi, K., Hoseney, R., Zeleznak, K. and Rogers, D. 1984. Effect of waxy barley starch and reheating on firmness of bread crumb. Cereal Chem. 61:281.

Guy, R. and Wren, J. 1968. A method for measuring the firmness of the cell-wall material of bread. Chem. Ind. (London). 1727.

Herz, K. 1965. Staling of bread - a review. Food Technol. 19:1828.

Hibberd, G. and Parker, N. 1985. Measurements of the compression properties of bread crumb. J. Texture Studies. 16:97.

Kai, T. 1985. Comparison of residual sugar and firming characteristics of white pan breads made by sponge dough and short time dough processes. M.S. Thesis, Kansas State University, Manhattan, KS.

Kilborn, R., Tipples, K. and Preston, K. 1982. The GRL compression tester: Description of the instrument and its application to the measurement of bread crumb properties. Cereal Chem. 60:134.

Kim, S. and D'Appolonia, B. 1977a. Bread staling studies. I. Effect of protein content on staling rate and bread crumb pasting properties. Cereal Chem. 54:207.

Kim, S. and D'Appolonia, B. 1977b. Bread staling studies. II. Effect of protein content and storage temperature on the role of starch. Cereal Chem. 54:216.

Kim, S. and D'Appolonia, B. 1977c. Bread staling studies. III. Effects of pentosans on dough

bread, and staling rate. Cereal Chem. 54:207.
Knightly, W. 1977. Bread staling - a review. Baker's Dig. 51:52.
Krog, N. and Davis, E. W. 1984. Starch-surfactant interaction related to bread staling - a review. Chapt. V-1 in: International Symposium on Advances in Baking Science and Technology. Tsen, C. C., ed. Dept. of Grain Science and Industry, Kansas State University, Manhattan, KS.
Kulp, K. 1975. Carbohydrases. Chapter 16 in: Enzymes in Food Processing. 2nd. ed. Reed, G. ed. Academic Press, New York.
Kulp, K. and Ponte, J. 1981. Staling of white pan bread: Fundamental causes. Crit. Rev. Food. Sci. Nutr. Vol. 15 (1):1.
Lasztity, R. 1980. Rheological studies on bread at the technical university of Budapest. J. Texture Studies. 11:81.
Maga, J. A. 1975. Bread Staling. Critical Rev. Food. Technol. 5:443.
Maleki, M., Hoseney, R. and Mattern, P. 1980. Effects of loaf volume, moisture content, and protein quality on the staling rate of bread. Cereal Chem. 57:136.
Marston, P. E. and Wannan, T. L. 1976. Bread baking, the transformation from dough to bread. Baker's Dig. 50:24.
McDermott, E. 1974. Measurement of the stickiness and other physical properties of bread crumb. J. Fd. Technol. 9:185.
Miller, B., Johnson, J. and Palmer, D. 1953. A comparison of cereal, fungal, and bacterial alpha-amylase as supplements for breadmaking. Food Technol. 7:38.
Moskowitz, H. R., Kapsalis, J. G., Cardello, A. V., Fishken, D., Maller, D. and Segles, R. A. 1979. Determining relationships among objective, expert and consumer measures of texture. Food. Technol. 33:84.
Neukom, H. and Markwalder, H. V. 1978. Oxidative gelation of wheat flour pentosans: a new way of cross-linking polymers. Cereal Foods World.

23:374.

Platt, W. 1940. Staling of bread. Cereal Chem. 7:1.

Ponte, J. G., Titcomb, S. T. and Cotton, R. H. 1962. Flour as a factor in bread firming. Cereal Chem. 39:437.

Prentice, N., Cuendet, L. S. and Geddes, W. F. 1954. Studies on bread staling. V. Effects of flour fractions and various starches on the firming of bread crumb. Cereal Chem. 31:188.

Russell, P. 1979. Current views on bread staling. Flour Milling Baking Res. Assoc. Bull. 122.

Russell, P. 1983. A kinetic study of bread staling by differential scanning calorimetry and compressibility measurements. The effect of different grists. J. Cereal Sci. 1:285.

Sandstedt, R., Schaumburg, L. and Fleming, J. 1954. The microscopic structure of bread and dough. Cereal Chem. 31:43.

Schoch, T. and French, D. 1947. Studies on bread staling. I. The role of starch. Cereal Chem. 24:321.

Schoch, T. J. 1965. Starch in bakery products. Baker's Dig. 39:48.

Shelton, D. R. and D'Appolonia, B. L. 1985. Carbohydrate functionality in the baking process. Cereal Foods World. 30:437.

Sidhu, J. S., Hoseney, R. C., Faubion, J. and Nordin, P. 1980. Reaction of 14C-cysteine with wheat flour water-solubles under ultraviolet light. Cereal Chem. 57:380.

Silberstein, O. 1964. Heat-stable bacterial alpha-amylase in baking-application to white bread. Baker's Dig. 38:66.

Varriano-Marston, E., Ke, V., Huang, G. and Ponte, J. 1980. Comparison of methods to determine starch gelatinization in bakery foods. Cereal Chem. 57:242.

Waldt, L. 1964. The problem of staling – its possible solution. Baker's Dig. 42:64.

Willhoft, E. 1971a. Bread Staling. I. Experimental study. J. Sci. Food Agric. 22:176.

Willhoft, E. 1971b. The theory and technique for measuring the firmness of a whole loaf by gaseous compression. J. Texture Studies. 2:296.

Willhoft. E. 1972. Moisture redistribution between gluten/starch interfaces in bread. IFST (U.K.) Broc. 5:67.

Zobel, H. 1973. A review of bread staling. Baker's Dig. 47:52.